The Water Revolution

Practical Solutions to Water Scarcity

Edited by Kendra Okonski

A publication of the Sustainable Development Network

Published by International Policy Press

The Water Revolution
Practical Solutions to Water Scarcity
Edited by Kendra Okonski

ISBN 1-905041-13-6

First published March 2006

International Policy Press
3rd Floor, Bedford Chambers
The Piazza, Covent Garden
London WC2E 8HA UK
info@policynetwork.net
www.policynetwork.net

t: +4420 7836 0750
f: +4420 7836 0756

Designed and typeset in Oranda by MacGuru Ltd
info@macguru.org.uk

Cover design by Sarah Hyndman

Printed in Great Britain by Hanway Print Centre
102–106 Essex Road
Islington N1 8LU

© 2006. Copyright of individual chapters retained by authors.

All rights reserved. Without limiting the rights under copyright reserved above, no part of this publication may be reproduced, stored or introduced into a retrieval system, or transmitted, in any form or by any means (electronic, mechanical, photocopying, recording or otherwise) without the prior written permission of both the copyright owner and the publisher of this book.

For questions regarding reprint permission and any other queries, please contact International Policy Press.

British Library Cataloguing in Publication Data. A catalogue record for this book is available from the British Library.

Contents

	Foreword *Hernando de Soto*	xi
	Foreword *Sir Ian Byatt*	xiv
	Introduction *Kendra Okonski*	1
1	**Comparing 20th-century trends in US and global agricultural water and land use** *Indur M. Goklany*	14
2	**Incentives matter: The case for market valuation of water** *Andrew P. Morriss*	38
3	**Reforming water policies in Latin America: Some lessons from Chile and Ecuador** *Douglas Southgate and Eugenio Figueroa B.*	73
4	**Poor provision of household water in India: How entrepreneurs respond to artificial scarcity** *Laveesh Bhandari and Aarti Khare*	93
5	**The rain catchers of Saurashtra, Gujarat** *Ambrish Mehta*	127
6	**Water governance in China: The failure of a top-down approach** *Wang Xinbo*	149
7	**The reality of water provision in urban Africa** *Franklin Cudjoe and Kendra Okonski*	179
8	**How not to reorganise an industry: Privatisation, liberalisation and Scottish water** *Colin Robinson*	204

Biographies

Laveesh Bhandari received his Ph.D. in Economics from Boston University in 1996. His Ph.D. dissertation was on the success and failure of international alliances. Prior to that, he earned an MA in Economics specializing in Finance and Industrial Organization, also from Boston University. His thesis received the "Best Thesis in International Economics" award by the EXIM Bank of India in 1996. He was also awarded the Hite Fellowship for his work on international finance. He has worked in the financial sector (Manhattan Funds, USA) where he was involved in the valuation of derivatives, in the development sector (National Council of Applied Economic Research, New Delhi) and now heads Indicus Analytics in New Delhi.

Dr Bhandari has extensively studied the Indian industry and infrastructure sectors as well as economic reforms. Apart from industry and infrastructure, he has worked on a wide range of topics such as poverty, nutrition, and the public distribution system, subcontracting, education and health of the poor, among many others. He has also taught economics at Boston University and IIT Delhi. He was also the Managing Editor of the *Journal of Emerging Market Finance*. Email: laveesh@indicus.net.

Franklin Cudjoe directs the Ghanaian think-tank Imani: The Centre for Humane Education, whose vision is to educate and create a core of young scholars that will promote market oriented policies throughout Africa. Prior to founding Imani, he was a programme officer and research assistant at the Institute of Economic Affairs in Ghana.

A Land Economist by training, Franklin works closely with partner think-tanks across the world to promote public policy ideas in Ghana and abroad. He is a frequent commentator in print and broadcast media about African development issues, including appearances on BBC, CBC, Swiss and Swedish National TV, Austrian National Radio and many Ghanaian media outlets. His work has been published in many newspapers around the world, including the *Ghanaian Daily Graphic*, *Accra Daily Mail*, London's *Daily Telegraph*, *The Wall Street Journal*, *El Mercurio* (Chile) and many others. He frequently speaks to policymakers, students groups in Ghana and to organizations abroad, and is an Adjunct Fellow at the Independent Institute in the USA. Email: franklin@imanighana.org

Eugenio Figueroa B. is Professor of Economics at the Department of Economics and Director of the Center for Environmental and Natural Resource Economics (CENRE) at the School of Economics and Business, and Executive Director of the National Center for the Environment (CENMA), at the University of Chile. He is also Adjunct Professor at the University of Alberta's Business School. He obtained his Ph.D. in Agricultural and Natural Resource Economics from the University of Maryland and his Masters in Economics from the University of Toronto. He also obtained his Doctorate in Veterinary Medicine (D.V.M.) and Bachelor in Animal Sciences from the University of Chile.

Professor Figueroa has taught and done research in several universities and academic institutions in America, Europe and Asia. He has published seven books, 30 chapters in different books and over 300 articles in scientific, academic and professional journals. He has also worked as economic advisor for governments and institutions in Africa, Asia, Europe and the Americas. He has conducted consulting projects for international organizations such as the World Bank, the Inter-American Development Bank, the United Nations and development agencies and companies in Canada, Germany, Japan, Spain, Sweden, the United States and Chile. Professor Figueroa has worked in several countries in the design, implementation and evaluation of public policy projects as they relate to natural resources, the environment, economic development, international trade, agricultural production and exports, and urban and rural development. Email: efiguero@cenma.cl

Dr. Indur M. Goklany is Assistant Director, Science & Technology

Policy, U.S. Department of the Interior. During his 30-plus years in federal and state governments, and the private sector, he has written over one hundred monographs, book chapters and papers on topics ranging from climate change, human well-being, economic development, technological change, and biotechnology to sustainable development.

Over half of his career has been with the U.S. Department of the Interior, which manages 20 percent of the U.S. land area, and associated mineral, energy and water resources. He was a visiting fellow with the American Enterprise Institute, and the Julian Simon Fellow at the Property and Environment Research Center in Bozeman, Montana. He has represented the U.S. at the Intergovernmental Panel on Climate Change and in the negotiations leading to the UN Framework Convention on Climate Change. His degrees, all in electrical engineering, are from the Indian Institute of Technology, Bombay, and Michigan State University.

He is the author of *Clearing the Air: The Real Story of the War on Air Pollution*, and *The Precautionary Principle*, both published by the Cato Institute. His next book, *Improving the State of Humanity and the Environment*, will be published in 2006. Opinions and views expressed by Dr. Goklany are his alone, and not necessarily of any institution with which he is associated.

Aarti Khare is an economist trained at Delhi School of Economics. She has conducted various studies on the Indian economy, business and markets. She has analyzed the performance of the Indian economy at the national, state and sub-state level. She is currently researching the market for fast-moving consumer goods in India. Email: aartikhare@yahoo.com.

Ambrish Mehta is a graduate in biology, and a senior member of an Action Research in Community Health and Development (ARCH), a non-governmental organization based in Gujarat, India. After graduation in 1980, he started working on the issue of proper rehabilitation of the tribal people affected by the Sardar Sarovar Project, a large multipurpose dam on the Narmada River. During this period, he also became interested in environmental issues relating to forests and water and started studying them.

He worked in the Gujarat Ecology Commission (GEC) from 1998 to 2002 as a nodal officer for the World Bank-assisted NGO Environment Action Fund. It was during this period that he first came in contact with 'rain catchers' of Saurashtra. While at the GEC, he also made significant contribution in the preparation of State Environment Action Plan for Gujarat, particularly in areas of forest and water management.

In 2001, he was selected as a Hubert Humphrey Fellow under the Fulbright Fellowship program. During his stay in United States, he studied in detail the intricacies of water rights as they have evolved in the western states of United States. He is currently the editor of 'Khoj' ('Pursuit'), a Gujarati magazine published by ARCH under its PAHEL (Initiative for Open Society) program.

Andrew P. Morriss is Galen J. Roush Professor of Business Law & Regulation at Case School of Law in Cleveland, Ohio; Senior Fellow, Property & Environment Research Center, Bozeman, Montana; and Senior Scholar, The Mercatus Center at George Mason University, Virginia. He received his Ph.D. in economics from MIT, his J.D. and M.Pub.Aff. from the University of Texas at Austin, and his A.B. from Princeton University.

He is the author or coauthor of more than forty book chapters and scholarly articles, including many on environmental law and policy, and is the co-author of *Regulation by Litigation*, forthcoming in 2007 from Yale University Press. He is also the co-editor of Cross-Border Human Resources, Labor and Employment Issues: Proceedings of the New York University 54th Annual Conference on Labor (with Samuel Estreicher) (Kluwer 2004), *Property Stories* (with Gerald Korngold) (Foundation Press, 2004); and *The Common Law and the Environment* (with Roger Meiners) (Rowman & Littlefield, 2000). His essays regularly appear in *The Freeman: Ideas on Liberty* and *Books & Culture: A Christian Review*. Email: andy.morriss@gmail.com

Kendra Okonski is Director of the Environment Programme at International Policy Network, a London-based development charity. She has worked for IPN since 2001. She is the editor or co-editor of several publications, including *Environment and Health* (2004) and *Adapt or Die* (2003), and an edition of the peer-reviewed journal *Energy and Environment*. She is also the author of *Montana: On the Verge of Collapse?* (Property and Environment Research Center, 2006). She frequently contributes to print and broadcast media on a realm of development and environment issues. She is a fellow of the Royal Society for the Arts and a Council Member of the University of Buckingham.

Prior to her work at IPN, she worked at the Competitive Enterprise Institute, a Washington, DC-based think tank, as a researcher to its president. She has a degree in economics from Hillsdale College. She is fluent in Spanish, and grew up in Chile and Montana. Email: kokonski@gmail.com

Colin Robinson was appointed in 1968 to the Chair of Economics at the University of Surrey. There he founded the Department of Economics and is now Emeritus Professor. From 1992 to 2002 he was Editorial Director of the Institute of Economic Affairs, a think tank in London.

His research is principally in the energy industries and the regulated utilities. He is the author of 23 books and monographs and over 150 journal papers. He was named British Institute of Energy Economics 'Economist of the Year' in 1992 and in 1998 received from the International Association for Energy Economics its 'Outstanding Contribution to the Profession and its Literature' award. Email: colin@gunnersbury.freeserve.co.uk.

Douglas Southgate is a natural resource economist with a Ph.D. from the University of Wisconsin, and has been a faculty member at Ohio State University since 1980. His research focuses mainly on environmental issues in the developing world, such as tropical deforestation and the economics of watershed management.

Along with numerous journal articles and scholarly papers, Dr. Southgate is the author of four books, including *The World Food Economy*, published in 2006 by Basil Blackwell. He also has consulted for the World Bank, the Inter-American Development Bank, the U.S. Agency for International Development, and the Ford Foundation in thirteen Latin American and Caribbean nations.

Wang Xinbo is associate professor at Capital University of Economy and Business (CUEB), Beijing, China, and co-director of the China Sustainable Development Research Center at CUEB. He is a PhD candidate in the graduate school of China Academy of Social Sciences (CASS), and earned an MA in economics in 1988. From 1988–1992, worked in the CASS Institute for Industrial Economics. From 2000–2004, he worked at the Unirule Institute of Economics in Beijing, and he is now an honorary research fellow of the institute.

His paper "Rethinking the nature of the firm" was published in China's most authoritative academic economics journal *JINGJI YANJIU* (Economic Research) in 1992, and it was republished in collection of papers on new institutional economics in 2003 by Beijing University Press. The paper was honoured for being an original contribution by a Chinese scholar to understanding in the discipline.

Recently, his academic research has focused on natural resources and environmental economics, especially from an institutional perspective. His working papers have included "A pricing mechanism for water supply projects" (2004) for a research program on water investment; "Urban water price formation and regulation" for the Chinese Ministry of Construction (2005); and "A preview of Beijing's market for reclaimed water" (2005). He is currently working as a consultant for JICA on China's water rights program. He frequently contributes articles to Chinese newspapers, and is fascinated by China's process of institutional transition.

Acknowledgements

We are grateful to the Earhart Foundation for supporting the production of *The Water Revolution* through a grant to International Policy Network.

The editor thanks all the contributors for their work and cooperation in the production of the book.

In addition, the book would not have been possible without the able research assistance of Caroline Boin and the editorial input of Mark Baillie, Julian Morris, Kristen Veblen and Linda Whetstone.

Reprint permissions

A version of Chapter 1, written by Indur M. Goklany, was published in *Water International*, Volume 27, Number 3, Pages 321–329, September 2002 and is reprinted here with permission of the International Water Resources Association.

Figure 3.1 in Chapter 3 was originally printed in Galiani, S., P. Gertler, and E. Schargrodsky (2005). "Water for Life: The Impact of the Privatization of Water Services on Child Mortality," *Journal of Political Economy*, vol. 113, pp. 83–120 (figure appeared on p.86). The figure is reprinted in this book with permission of the University of Chicago Press.

A version of Chapter 8, written by Colin Robinson, was published in 2005 by the Policy Institute, Edinburgh, Scotland, and is reprinted here with permission, with thanks to Tom Miers, Director of the Policy Institute.

Foreword

For decades now I have been arguing that the poor are not the problem but a big part of the solution. The poor have assets – houses, land, livestock, and businesses. Tragically, those assets remain dead capital. Hostile, discriminatory, and costly legal systems throughout the developing and post-Soviet world have excluded the poor – four billion of the world's six billion people – from the very legal tools that would allow them to leverage their assets: forms of organizing their enterprises that would allow them to divide labor and increase productivity, identity devices that would give them access to markets outside the confines of their families and acquaintances, and legal property rights that allow their assets to be given different economic functions so as to capture the highest economic value in the market.

Legal property is not simply about ownership; it is a basic right that gives people access to the courts, to banks, to a way of settling disputes without violence; property rights not only increase access to credit and capital formation, they provide individual accountability, facilitate enforcement, improve governance, reduce corruption, fight terrorism; legal property creates incentives for investment, making it more profitable for the private sector to provide roads, electricity, and water to poor areas, develop affordable housing, and set up information systems that assist the poor. In short, any government serious about grappling with one of the world's most persistent and destabilizing problems – widespread, desperate poverty – must give all its citizens easy access to legal property.

This important book shows that extending legal property rights to poor people will also help solve another of the 21^{St} Century's most pressing problems: water scarcity. Ominous signs are already staring us in the face. All the countries of Central Asia, the Middle East, and North Africa are facing water deficits. Much of urban India does not have 24-hour water supplies; half of China's 600 cities faced water shortages throughout the 1990's.

For most Americans, it is unimaginable that the great Mississippi River would one day begin to dry up and not reach the ocean. Yet between 1974 and 2000, China's Yellow River – 556 kilometers shorter than the Mississippi – ran dry 18 times. In 1998, the Yellow River failed to reach its ocean mouth for more than 250 days. With 1.3 billion people to feed, such water shortages are not just a major agricultural problem but a serious threat to China's economic and political stability.

Scarce water means expensive water. Rich countries, such as all of the oil-producing Middle Eastern states where aquifers are being pumped dry, will be able to cope, at least in the short term. For poor countries in Africa and Latin America which are already hard-pressed to develop their economies and safeguard their toddler democracies, water scarcity threatens not just the health of their people and the environment, but peace itself. The poor are already angry that the economic benefits of the market reforms of the past 16 years have not trickled down their way. What will they do when the price of water – and food – increases because of water shortages?

The successful nations of the world, led by the U.S., are quick to sing the praises of "market forces" – while ignoring the fact that four billion people in the world have been shut out of the market. The market is not the privilege of the rich. It is the fundamental thread in the social fabric of any society. Bad laws and inefficient bureaucracies, however, impose high transaction costs that preclude ordinary people from accessing the basic institutions that would empower them as full members of a market economy. Shut out of the legal economy, the poor create their own, extralegal economy and their own rules and regulations to make it work.

These local, grassroots practices have been discovered and documented by my colleagues at the Institute of Liberty and Democracy in Lima, Peru, through our projects in the underground economies of Latin America, Asia, the Middle East, and Africa, where majority of people are poor and operate outside the law.

In Tanzania, for example, we recently found that 98 percent of the businesses and 89 percent of the property are held extra-legally

– an estimated $29 billion worth of dead capital in a country with a per capita GDP of $210. But the ILD also found that to make a living and protect their assets, ordinary Tanzanians had created their own property rights, documents, registries, identification systems, business associations, wills, transparent and standardized accounting practices. They have invented ways to divide labor, build management, create collateral, and trace liability. In short, Tanzania already has in place the building blocks for a modern economy. The challenge, as it is throughout the developing world, is to integrate the people's law into a modern, inclusive economic system under one rule of law that gives everyone a stake in the market.

And then step back and watch the people work and innovate. In this book, a group of highly regarded experts argue impressively and illustrate in many different ways that if poor entrepreneurs are allowed to flourish unhindered by bureaucratic intervention and fueled by better laws, they will come up with ways of providing and conserving water.

Chapters 4 and 7, for example, discuss the successes of "informal" and "illegal" water markets in India and Africa in providing water in shantytowns. Chapter 6 offers the leaders of China a way out of those dry riverbeds, but only by means of bottom-up solutions rooted in local property rights and water user associations. *The Water Revolution*'s other authors argue that well-intentioned government efforts to protect water supplies perpetuate artificial scarcity and often result in wasting water; private, market oriented solutions are often more innovative and productive.

The policy implications are as clear as a glass of sparkling clean water: Instead of generating legal obstacles that keep the poor on the outside looking in, nations must legally empower the poor so that they can pull themselves out of poverty – and, as the essays in *The Water Revolution* argue so forcefully, they must explore where it makes sense to use market solutions to address the growing problem of water scarcity all around the world.

Hernando de Soto
President of the Institute for Liberty and Democracy based in
Lima, Peru and author of *The Mystery of Capital*
February 2006

Foreword

Governments rarely manage water supplies well and often bear the responsibility for shortages and inferior quality. As a practitioner, I greatly welcome this book, which demonstrates how markets can work to the public good.

Experience has shown that markets produce good outcomes for users. The task for the public authorities is not to create or preserve an inefficient state monopoly, but to identify where competition is possible and to institute incentive regulation where it is not.

The UK experience shows that regulation can be designed to provide incentives to suppliers and users. Privatisation of water services has led to a doubling of investment, substantial improvement in services to customers and much greater efficiency among suppliers. Although there are no public subsidies, tariffs have risen only modestly. The market is now open to competition for larger business customers in England and Wales and the retail market for all business customers in Scotland. This should lead to further benefits for customers.

The studies in this book show the crucial importance of involving people rather than politicians in water.

Indur Goklany documents how agriculture has relied on greater amounts of water in irrigation for food production, and argues for greater reliance on property rights and the avoidance of subsidies properly to manage water resources.

Andrew Morriss argues that markets for water hold the potential to unleash significant entrepreneurial activity in water production and consumption.

Douglas Southgate and Eugenio Figueroa document the differences between Ecuador, where subsidies for potable water and for irrigation have been damaging, and Chile, where the country's water law has allowed market-based allocation of scarce water.

Laveesh Bhandari and Aarti Khare argue that in urban India, government suppliers have created an artificial scarcity and that the conditions for private sector water provision do not require micromanagement or micro-regulation but would be facilitated by simple broad policy measures.

Ambrish Mehta explains how farmers and the broader public in one region of Gujarat, India, cooperated effectively to collect rainwater and improve crops despite hindrance rather than help from government.

Wang Xinbo illustrates the environmental and investment problems in China caused by its top-down, central planning and argues that China's sustainable development will require a decentralized approach.

Franklin Cudjoe and Kendra Okonski show that in Africa's urban areas today, the poor have been neglected by their inept, corrupt governments, which have failed to deliver water and sanitation. The solution, they argue, is to enable 'entrepreneurs of all sizes, including poor informal entrepreneurs operating at the lowest level of society, to deliver water and sanitation services to their fellow citizens'.

Finally, Colin Robinson points out the dangers of relying exclusively on a state corporation in Scotland, which presents the likelihood of political interference. He emphasises the need clearly to separate the activities that could be competitive from the monopoly network.

Subsidies create waste. Active customers, rather than passive consumers, will develop their own solutions, especially in rural areas. The operation of public networks can be dramatically improved where there is proper, non-political oversight of service delivery. If privatisation of public networks is to produce a lasting solution, it must be well designed. It should not be undertaken solely to help the public finances. Experience has shown, however, that if regulators are given the right duties and the right powers they can play a major role in developing practical forms of competition in network businesses.

Ian Byatt
February 2006

Executive summary

Chapter 1

Comparing 20th century trends in US and global agricultural water and land use

- During the 20th century, huge gains in the efficiency of use of cropland for food and fibre were achieved.
- During the same period, similar gains were not achieved in the efficiency of use of water.
- The disparity can be explained largely by differences in the degree to which land and water were subject to private ownership.
- Most cropland was privately owned; whereas most water was state property.
- Private ownership encourage responsible and intelligent use of resources, leading to improved agricultural productivity and environmental conservation.

Chapter 2

Incentives matter

- For private ownership to be effective, property rights must be "3D":
  - definable (capable of clear definition),
  - defensible (capable of being defended in courts of law), and
  - defeasible (capable of being transferred to others)
- 3D property rights enable adaptation to dynamic changes through market processes.
- Markets enable people to discover the value of water of differing quality at different times and in different places.
- Markets are a superior way to generate and transmit information, at a low cost, thereby enabling greater productivity and better satisfaction of people's needs and wants.

Executive summary xvii

- Competition in water markets creates incentives for private companies and individuals to respond to changing supply and demand by developing new technologies and utilising water more efficiently.

Chapter 3

Reforming water policies in Latin America
- A case study of urban water provision in Quito, Ecuador, shows that public management of water tends to be inequitable and unsustainable.
- Until recently poor people in Quito are not well served by the municipal water supplier.
- Meanwhile, subsidised irrigation throughout Ecuador benefits well-off farmers but perpetuates environmental problems.
- The poor are harmed most by the government's inadequate recovery of investments in water provision.
- The dire situation in Ecuador is contrasted with that of Chile, which enacted a property-rights-based water policy in the 1980s.
- Underpinned by legally secure and transferable property rights, Chile's system has achieved nearly universal access to water for both urban and rural users, including the poor.
- Farmers in Chile have benefited immensely from being able to trade water.

Chapter 4

Poor provision of household water in urban India
- India is blessed with more than adequate hydrological resources, but there is an artificial water shortage in urban areas.
- Since water is practically free, public sector providers have no incentive to prevent waste or ameliorate existing infrastructure.

- A case study of an illegal squatter settlement ("basti") in New Delhi shows that informal sector entrepreneurs are solving water scarcity for poorer urban residents. Compared to the public sector, they service the poor more efficiently, more conveniently and more widely.
- India's water still needs to be used more efficiently. Enabling the private sector to process, transmit and distribute water would improve human health and enable investments in infrastructure.

Chapter 5

The rain catchers of Saurashtra, Gujarat

- The peninsula of Saurashtra, Gujarat, is a region whose geographical conditions create acute water scarcity.
- Government solutions in the form of reservoirs have largely benefited urban, but not rural, populations.
- Farmers and others have created more water, for themselves and others, with water harvesting devices such as check-dams in streams and rivers, well recharging, and other methods of "rain catching."
- The lesson for other parts of India is that informal, decentralized arrangements can play a major role in solving water scarcity, even if the outside world perceives them to be chaotic and disorganized.

Chapter 6

Water governance in China

- China's water governance has been characterised by a "top-down" approach.
- This has led to huge inefficiencies in water use and extensive water pollution.
- For instance, the Huaihe River remains polluted in spite of expensive clean-up efforts, while government agencies were

Executive summary xix

incompetently handled a recent pollution incident on the Songari River.
- The solution to China's water scarcity is to establish a legal framework that respects individual property rights.
- In combination with other local arrangements, this would enable and lead to respect for individual and community decision-making.

Chapter 7

The reality of water provision in urban Africa
- Across Africa, governments largely have deemed that water provision is the responsibility of the state.
- In practice, public sector water systems suffer huge losses, are politically driven and largely have been incapable of extending the network to peripheral urban areas.
- Some aspects of privatization in Senegal, Cote d'Ivoire and Guinea successfully resolved such problems.
- In many urban areas – especially areas not formally recognized by government planning systems – informal ("illegal") entrepreneurs are helping to solve the water and sanitation gap created by governments.
- Inadequate institutions prevent entrepreneurs of all sizes from delivering water and sanitation services through markets. Poorer Africans suffer most in this environment.
- Reforms to extend formal property rights and other institutions that support markets would both solve artificial water scarcity and enable economic growth more generally.

Chapter 8

How not to reorganise an industry
- National government ownership of goods and service industries leads to inferior outcomes compared to private ownership with competitive markets.

- The 1989 privatisation of the water and sewerage industry in England and Wales has produced enormous benefits for consumers, although it still has flaws.
- However, Scottish Water remains nationally owned, with millions of captive customers and little possibility for competition.
- Recent reforms may help to improve the service offered by Scottish Water, but the only way genuinely to ensure efficiency, quality and sufficient investment is a combination of privatisation and market liberalisation.

Introduction: Why markets matter

Kendra Okonski

Kendra, the editor of this book, lives on the outskirts of London. After water privatisation in England and Wales in 1989, water and sewerage services have been provided to households by private companies operating essentially as regulated local monopolies. One of these private companies supplies her water.

Despite its reputation for an extremely wet climate, Southeast England has been unusually dry in recent years. This has presented the possibility that people might need to use water more efficiently.

For households fully to understand – and hence manage – their water consumption, their water use must be metered and each drop subject to a charge. In the summer of 2005, a water company representative visited her residence, determined the need for a meter, and marked an area outside for its installation. Amidst the possibility of a more intense drought in the summer of 2006, she is still waiting for the meter seven months later.

If this situation is multiplied across millions of households in London and Southeast England, then individual contributions to averting a potential drought may not be forthcoming. Some estimates suggest that London's water system loses 40 percent of its water (a problem resulting from relatively older infrastructure). As such, it could make sense for water companies to invest in updates and improvements to ensure that less water is lost.

However, if water companies do not possess a means to measures water consumption by individual households, then they lack sufficient information about how prices should reflect the relative scarcity of water. Thus, they will not know how much to invest in improving and repairing infrastructure.

An equally important issue presented by the potential drought is the economic implications of excessive water being used for subsidised agriculture in the region. If prices accurately reflected relative scarcity, it is unclear that this economic activity would be a viable use of water.

Issues raised in the context of a potential drought in Southeast England may seem petty compared to water issues in other regions around the world. Indeed, these problems pale in comparison to grave issues pertaining to water in poorer countries considered extensively by contributors to this book.

However, the greater wealth of people in the UK and other rich countries enables us to be concerned about the individual and aggregate implications of our water use. Indeed, we might call this a consequence of the 'environmental transition', an idea discussed by Indur Goklany (Chapter 1). Since we have already addressed a more urgent and fundamental problem – having access to a clean, reliable supply of water – we can invest scarce resources in ensuring that we have an ever-more sustainable supply of water.

With that in mind, contributors to this book address a wide range of geographical areas and topics. They consider urban water and sanitation, and also examine the use of water in agriculture and industry more generally. They offer both theoretical explanations and wisdom derived from practical examples. The chapters are not intended as a technical manual, and they are intended to provide a representative rather than exhaustive perspective.

A 'common good'?

It is often claimed that water is a "common good." This observation is true to some extent: water is, indeed, all around us. It falls from the sky into oceans, lakes, rivers, and streams, in different proportions around the planet. The hydrological cycle is itself an amazing process by which water constantly is recycled in the earth's atmosphere. As a result, some regions are blessed naturally with sufficient water (and sometimes too much), while other areas receive relatively less.

However, the claim that water is a 'common good' is often a ruse for justifying all manner of inappropriate policies for its use and management. These include – but are not limited to – government subsidies both to companies and users (for which taxpayers ultimately foot the bill); allocation by politicians and government agencies; and collective ownership. Ownership of water is truly a complex issue relating to the presence or absence of institutional arrangements in any society at a given time, discussed in more detail below.

Referring to water as a "common good" is also used to justify opposition to any form of valuation by commercial or individual means, through prices (but more importantly, through markets which generate prices). Usually without any more discussion, such solutions automatically are castigated as being greedy, selfish, profiteering, a symptom of our "ever-more-commercialised culture", and a derision of our "common heritage."

Those who repeat the notion that water is a "common good" seem not to understand that water in its natural form is usually not appropriate for human uses. For instance, humans are ill-advised to drink water directly obtained from streams, lakes and rivers because it is likely to contain bacteria which would make us sick. In fact, somewhere in the range of two to four million people (including at least 1.7 million children) die every year around the world because they contract diarrhoea and other diseases from water which has not been sufficiently treated and processed (WHO 2005).

At the same time, the scientific and industrial procedures entailed in the production of medicines to treat many kinds of diseases (including new vaccines to prevent diarrhoea) require water of nearly 100 percent purity. In its natural state, water is not clean or pure enough for such uses.

Water is both a vital and instrumental good. It is vital because all life requires water. It is instrumental because humans use water in a variety of products and processes, at different times, in different places and with different quality requirements.

Humans in both urban and rural areas need to use water, but many of these uses are quite different. Dense urban areas yield many benefits – but delivery of clean water and removal of dirty water (the inevitable by-product of human settlement) becomes especially complex when people live in concentrated settings.

In contrast, rural areas may or may not have access to groundwater for household use, and poor sanitation may present fewer hazards than in an urban setting. However, rural people in poor countries around the world largely are engaged in agricultural production. A majority of the world's water is used in irrigation, and the availability of water can make or break a poor farmer.

The process of delivering a specific quality of water to the right place, and at the right time, involves expenditure of resources. Potable water, sewerage and wastewater treatment systems require pipes, treatment facilities and proper management (in the form of skills that are in relatively short supply). When such systems do not exist, people must spend time and resources to acquire clean water and to dispose of dirty water. Irrigation water must be provided by someone and must come from somewhere – be it surface water in reservoirs, groundwater or other sources – and this involves delivery costs.

In short, there are economic costs entailed in processing, delivering and removing water in all of its embodiments, regardless of who performs these services. This is why water is an economic good which deserves to be included in, rather than excluded from markets. To encourage the best use of scarce water, it is imperative that humans be able to value water in its different uses: in a potable form piped to households, as a service used for environmental purposes, and used to produce goods and services.

The contributors to *The Water Revolution* present considerable evidence to support the view that markets, especially in the context of supporting institutions, and especially when compared to alternatives, produce the fairest and least discriminatory outcomes, the most innovation, the most environmental benefits and the most efficient and dynamic uses of scarce water.

The role of markets

Colin Robinson (Chapter 8) – observes that markets, like all human institutions, work "imperfectly." Thus, it is not helpful to use "perfection" as the standard by which we evaluate water services (Segerfeldt 2005). It is useful, for our purposes, to consider degrees of "imperfection." Judging from contributions to this book, government allocation of water wins hands-down.

Based on two paradigmatic examples of countries (Chile and Ecuador) which have followed two different paths for managing water, Douglas Southgate and Eugenio Figueroa argue that non-market allocation frequently "creates inefficiency, inequity and damage to the environment" (Chapter 3).

First, they show that in both urban and rural areas, poor people "derive little benefit from subsidies – including poor cost-recovery in potable-water systems." During the 1980s, Quito's municipal water company recovered only 50 percent of its costs. Similar losses are experienced in urban water systems of many African countries (Chapter 7).

Why do economists refer to poor "cost-recovery"? Costs matter, because they reflect the relative scarcity of a resource. Infrastructure, such as treatment facilities and pipes, entails a cost – because the materials in those pipes could be used to produce other goods. Maintaining that infrastructure has a cost – because people expect to be compensated for the use of their skills (most people tend to be averse to working "for free", all of the time). A water system may not cover its costs if it loses large or excessive amounts of water, if it subsidises consumption for industry and households, or if it does not bill its users and collect their payments. Fundamentally, poor cost-recovery means that a system does not generate sufficient revenue to invest in maintenance, innovation, or environmental improvements, such as watershed conservation.

The crucial difference between public and private sector systems is that private sector water providers possess a metric – in the form of market-driven prices –which enable them to measure and control their costs, in the interest of earning a profit.

Colin Robinson (Chapter 8) discusses the theoretical and practical aspects of public versus private sector provision of water. He contrasts privatised water and sewerage companies in England and Wales with Scotland's nationalised water company, which has millions of "captive customers." Nationalised industries in England and Wales before privatisation had poor standards of customer service – but customers had no other choice. Robinson concludes that "almost always and everywhere, politicians' bans on competition are an extremely bad idea."

Those who maintain that water is managed best when it is owned collectively should pay careful attention to the perilous situation of China (Chapter 6). The state has declared that it owns all of the country's water and until recently, the state allocated licenses to water users. In a country where water is already relatively scarce, this "top-down" management has led to hugely inefficient utilization, extensive water pollution and adverse human health impacts.

Wang Xinbo observes that China's shift towards private sector investment in water supply (particularly in urban areas) has been a difficult process because it lacks pre-existing market disciplines. The country's public utilities are overstaffed but have poor maintenance (both are characteristic of government provision of water). The utilities have accounting methods used by China's government agencies – and lack a basic tally of costs and assets. The utilities have also overestimated water supplies, and have thus overinvested in capacity, a primary cause of China's "pump race."

What all of this demonstrates is that subjecting water providers and users to the discipline of competitive market processes is the best way to ensure both that costs do not spiral out of control, and that customers receive quality service at competitive prices. A company which does not pay attention to those critical details is liable to go out of business (unless it receives a government subsidy).

It is often alleged that water is deserving of taxpayer subsidies, in one form or another – whether the subsidy accrues to poor household users, farmers or industry. In Ecuador, subsidised irrigation generally benefits the politically-connected and privileged elite and not the rural poor (Chapter 3). The same story is true in most countries. The elite accrue the benefits of subsidies through increased values of real estate – while the costs are dispersed among the tax-paying public.

Poor households in African cities are also unlikely to benefit from subsidised water (Chapter 7) for a variety of reasons. One is that municipal governments require proof of property ownership to grant access to subsidised connections; it is truly perverse that governments deny their poorest citizens the ability legally to own their dwellings. Another is that government-sanctioned vendors are liable to mark up the price of subsidised water provided by a utility: in the case of Nairobi, Kenya, the price was 18 times greater.

The belief that governments will allocate water efficiently and fairly is based on the flawed assumption that public authorities will "faithfully safeguard the public interest by turning a deaf ear" to rent-seekers (Chapter 5). In practice, this is not the case, whether in China, India, Ecuador, Kenya, Ghana, Tanzania or the UK.

The reason is not, as Colin Robinson points out (Chapter 8), that government officials are bad people. It is because politicians, government authorities and public officials *are* people: their incentives do not change in a political setting.

This is why politicians fall victim to 'rent-seeking': their political power enables them to allocate government's resources to themselves, their friends and lobbying interest groups. Across the world, political interest groups are concentrated and powerful, and they respond to incentives created by the political system. In the world's poor countries being a member of the wealthy elite is often synonymous with possessing political power – and vice-versa.

This helps to explain why across Africa, government officials are averse to recognising the existence of informal settlements where new urban populations live, such as shanty-towns and slums

(Chapter 7). By denying these people land tenure and refusing to extend public services such as water, the implication seems to be that if they are ignored, they will go away.

Both municipal and national governments have contributed to the water and sanitation gap in African countries. Because they have no metric (prices and costs deriving from markets) by which to decide whether or not to invest in extending their water and sewerage networks, they view extra people as a burden on systems which are already strapped for cash. But there is a way out of this mire, both literally and figuratively.

First, it is imperative that urban planning systems in African countries are reformed so that they formally recognise their growing populations. Currently, the definition of an urban area is the area which benefits from services ostensibly provided by governments – such as water and electricity. This is tautological and counterproductive.

Second, entrepreneurs – of all sizes, shapes and forms – view extra people as a business opportunity. In Africa's cities, they have found cost-effective ways to deliver water and provide sewerage services. Informal entrepreneurs supply those services to fellow residents of slums and shanty-towns (Chapter 7). Albeit an unenviable job, they supply a necessary service in exchange for payment. A similar situation exists in illegal squatter settlements in urban India (Chapter 4), where residents pay to have piped water delivered to their homes twice daily.

At the same time, governments in India and Africa perpetuate barriers to entrepreneurship which can be referred to as "transaction costs." These transaction costs are especially harmful for the operations of informal sector entrepreneurs. It potentially could take months, if not years, and dozens of procedures to obtain legal recognition for a business, or to enforce a contract.

Though competition between informal entrepreneurs is relatively free, the fact that governments have not formally recognised their existence is extremely problematic.

Without formal sanction, these entrepreneurs are held back. It is prohibitively expensive for them to expand their services, to innovate and address problems. While they can acquire capital from their families and friends, they cannot obtain larger loans from commercial lenders. Moreover, they are unlikely to accrue all of the benefits of their own innovation and investment.

In part these problems exist because the state may apply its blunt and heavy hand if given the opportunity – whether this entails extortion or confiscation of capital. High transaction costs mean that those entrepreneurs are unable to take advantage of potential economies of scale. Reforms are imperative both to achieve better allocation of water, and to enable economic development more generally.

As demonstrated so well in the case of Chile (Chapter 3), this strategy can yield huge benefits for the poor and for the environment. Douglas Southgate and Eugenio Figueroa doubt that "the competition over water resources inevitably created by economic expansion could have been resolved as effectively in the absence of policies that stress ownership and markets."

A fundamental lesson of this volume is that the right kind of privatization entails the creation of an enabling environment for entrepreneurs. To harness the full power of human initiative and market competition requires the market process to be underpinned by supporting institutions. Andrew Morriss discusses the nature of these institutions (Chapter 2), noting that they must be flexible enough to accommodate the dynamic nature of water uses.

Markets, innovation and the environment

Indur Goklany (Chapter 1) demonstrates that during the 20th Century, the benefits of technological innovation accrued to land, but largely not to water. Land use became more efficient because of investments that enabled better crop yields, and more efficient use of agricultural inputs such as pesticides and fertilizers. This process yielded many tangible benefits to humanity and the environment. Goklany argues convincingly that it is the existence of property rights and markets which explains relative gains in efficiency in land, compared to parallel inefficiencies in water.

It is here that the true benefits of markets are apparent. In the absence of markets, argues Andrew Morriss (Chapter 2), users of water have no way to determine the value of their scarce resources. This prevents beneficial trades from occurring.

Chile is one example where reforms enabling the use of markets to allocate water have simultaneously created economic and environmental benefits (Chapter 3). The Limarí Valley, an arid area north of the capital city, Santiago, has a spot market for water which has reduced the costs of transactions between farmers. Farmers can fully understand its relative value in different uses. Because they pay the full cost of their water, they have an incentive not to waste it. As a result, they produce crops with a higher economic value, such as grapes for wine production and fruit for export.

The same phenomenon has occurred in Gujarat, India, with groundwater, which is managed by users and riparian owners (Chapter 5). More and more, farmers here are investing in devices such as drips and sprinklers to enable them to squeeze the most out of every drop of water. As a result, they have invested in producing higher-valued orchard crops.

All of this provides solid evidence to corroborate the insights of both Indur Goklany (Chapter 1) and Andrew Morriss (Chapter 2). Morriss explains that using markets to allocate water creates superior outcomes in light of alternatives, because they enable people to transact with each other. By enabling more transactions to occur, markets reduce the costs of transacting. Markets generate a great deal more information compared to alternative arrangements, which in turn feeds innovation. One consequence of innovation is that more may be achieved with fewer resources; in the case of water, this means developing technologies that enable conservation, re-use or recycling. Such innovation could apply across the board in household and industrial uses of water.

Those who oppose market allocation of water have suggested that environmental resources should be owned collectively. Similarly, they portend that extending market institutions to the environment will result in certain gloom and doom. Practical experience shows us why this is not true. Insofar as resources are owned collectively or even by governments, in practice they are owned by no one. Andrew Morriss points to the tragedy of the Aral Sea, "perhaps the largest environmental disaster relating to water in modern times." He notes that "only a government can create a disaster of such a magnitude, for only a government can seize property rights on such a scale without paying compensation."

Allowing institutional arrangements to evolve

Ambrish Mehta (Chapter 5) provides a case study which explains how farmers in one region of India used decentralized initiatives to address water scarcity. Outside "experts" claim that their methods are chaotic, on the basis that they have not been subjected to top-down government planning.

He demonstrates clearly and eloquently why local people – not the collective mass of humanity as represented by the government of Gujarat, the government of India or the United Nations – are better at protecting the environment. Local people are more likely to be intimately involved with a resource. They develop institutional arrangements that enable their resources to be utilised and managed in a specific time and place, but those complexities may not be apparent to the naked eye. Farmers in Saurashtra have identified and innovated a solution to manage local water. This solution has "expanded the pie" (in terms of water availability) rather than creating conflict over an "existing pie."

Saurashtra's farmers have re-asserted their riparian rights in the face of government intervention. Their institutional arrangement has evolved in their specific context, but it might not work elsewhere. Other parts of India have evolved similar solutions to encompass local knowledge (Shah 2005). Importing particular "solutions" from elsewhere or imposing them with top-down governance is unlikely to create long-term, peaceful solutions to water (and resource) scarcity.

A parallel example, which is more formalized, involves protection of water courses in England and Wales. As documented by Roger Bate (2003), the Anglers Conservation Association (ACA) uses common law to defend and protect water courses from polluters. This not only improves water quality for fish and anglers, but also creates environmental amenities for additional users. Because the ACA relies on formal property rights, it has even been successful at challenging pollution caused by the state.

Conclusions

A degree, or even a large dose, of pragmatism is needed in the debate about water. Solving water scarcity in the 21st Century means extending – not narrowing – the role of markets and their underlying institutions. The poor, the rest of humanity and the environment will benefit from practical solutions entailed by market solutions to water scarcity. Water is vital and deserves to receive the full benefits of being subjected to competitive market processes.

Markets and their supporting institutions are one of the best means we humans have for dealing with one another in utilizing and managing our scarce resources. This is not to say that such arrangements are the "end all, be all" of human existence. Instead, we should view markets as an instrumental good which enable humans to fulfil our myriad needs and pursue our myriad goals. In an era where the people in the world are becoming ever more connected to each other, markets and their underlying institutions must be strengthened.

References

Bate, Roger. (2003). "Saving Our Streams: The Role of the Anglers' Conservation Association in Protecting English and Welsh Rivers." *Fordham Environmental Law Review*, vol.14, no.2 (Spring).

Segerfeldt, Fredrik. (2005). *Water for Sale: How Business and the Market can Resolve the World's Water Crisis*. Washington, DC: Cato Institute.

Shah, Tushaar (2005). "The New Institutional Economics of India's Water Policy." Conference paper presented at "African Water Laws", 26–28 January, Johannesburg, South Africa. Online: http://www.nri.org/waterlaw/AWLworkshop/SHAH-T.pdf. Visited 14 February 2006.

World Health Organization (WHO)(2005). *World Health Report*. Statistical Annex. Online: http://www.who.int/whr/2005/annexes-en.pdf. Cited 14 February 2006.

1 Comparing 20th-century trends in US and global agricultural water and land use

Indur M. Goklany

Nearly everywhere in the world where humans use land and water, they primarily use it for agriculture. The amount of land used per capita for agriculture has been in decline since the early 1900s. However, agricultural water use per capita only began to decline in the latter half of the 20th Century. One factor that helps explain these facts is that farmers (and farming communities) have traditionally had stronger property rights to their land than to their water. As a result, through much of the 20th Century, farmers had a greater incentive to improve the efficiency of land use than that of water use, and to substitute water for land (or irrigated land for dryland) in producing crops.

Land and water are the two most critical natural resource inputs for agriculture. Globally, agriculture accounts for 38 percent of land use, 66 percent of freshwater withdrawals, and 85 percent of freshwater consumption (FAO 2001; Shiklomanov 2000).

Agriculture inevitably has a significant impact on terrestrial and freshwater habitats, ecosystems, and biological diversity (Wilson 1992; Goklany 1998, 1999a; IUCN 2000). It is generally recognised that conversion of land to agricultural uses is the single most important threat to terrestrial biodiversity. According to the International Union for the Conservation of Nature, habitat loss and degradation – to which agriculture is a major contributor – affect 89 percent of

Figure 1.1 **US cropland and irrigation water use 1910–2004**

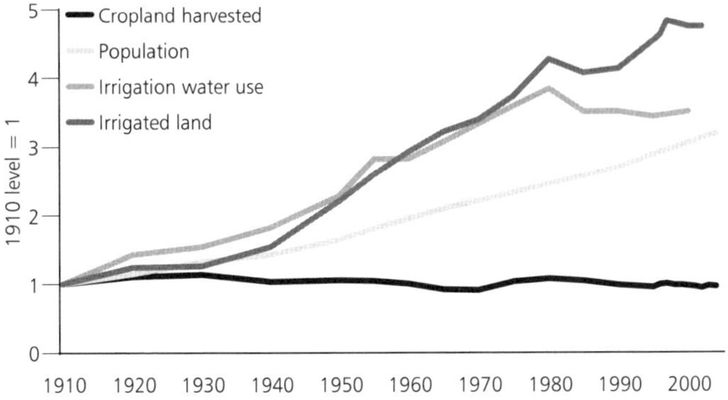

Sources:
Irrigated land: from 1910–1955, USBOC (1975, p. 433); from 1960–1995, USDA (2001), p. ix–7 (interpolated, as necessary); for 2000, Gollehon et al. (2003);
Irrigation water: USBOC (1975), p. 434; for 1950–2000, Solley et al. (1998) and Hutson et al. (2005);
Cropland: for 1910–1995, USDA (no date); for 1996–2004, USDA (2005) Agricultural Statistics 2005, p. ix–17.]
Population: Bureau of the Census (USBOC)
(1) 2000–2004: 'National and State Population Estimates: Annual Population Estimates 2000 to 2004,' at http://www.census.gov/popest/states/NST-ann-est.html visited August 14, 2005.
(2) 1900–1959: Historical National Population Estimates, 1900 to 1999 at http://www.census.gov/popest/archives/pre-1980/ visited August 14, 2005. (3) 1960–1999: Statistical Abstract 2004–2005

birds, 83 percent of mammals, and 91 percent of plants assessed by the organization to be 'threatened' (IUCN 2000).

Similarly, water diversions for agricultural uses and the pollution generated by agricultural practices contribute significantly to the threats facing many freshwater species (IUCN 2000; Wilson 1992). Although robust information and data are lacking, it is estimated that about 20 percent of freshwater species are threatened, endangered, or extinct due to a variety of causes, including agricultural demand (IUCN 1999).

This chapter considers the trends in water and land use over the course of the past 100 years, in the US and globally, and the factors underlying these trends.

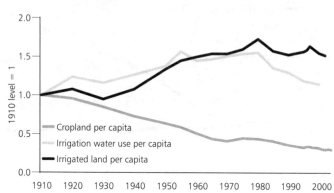

Figure 1.2 **US per capita cropland and irrigation water use 1910–2004**

Sources: USBOC (1975, various years), Solley et al. (1998), Hutson et al. (2005), Gollehon et al. (2003), USDA (2001, 2005).

US trends: 1910 to 2004

In the US, agriculture currently accounts for one-third of surface water withdrawals, two-thirds of groundwater withdrawals, and 85 percent of consumptive water use (Solley et al. 1998). Meanwhile, harvested cropland accounts for 16 percent of US land area excluding Alaska (US Bureau of the Census 2006; US Department of Agriculture 2001a).[1]

Between 1910 and 2000, the US population increased by 205 percent. Despite the increase in demand for food, the amount of cropland harvested declined by 3 percent. However, as can be seen from Figure 1.1, total water withdrawn and used for irrigation increased by 251 percent (US Bureau of the Census, 1975; Solley et al. 1998; US Department of Agriculture 2001b, 2005a, 2005b; Gollehon et al. 2003; Hutson et al. 2005)). Meanwhile, the amount of irrigated land increased by 374 percent.[2] Over the same period, yields per unit of land increased substantially for

Comparing 20th-century trends in agricultural water and land use 17

Figure 1.3 **Global cropland and irrigation water use 1900–2000**

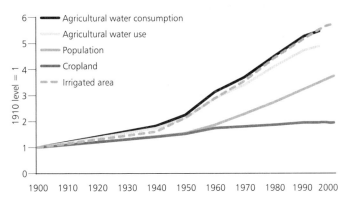

Sources: for water and irrigated land, Shiklomanov (2000); for population, McEvedy and Jones (1978) and FAO (2002); for cropland, Goklany (1999a) and FAO (2002). Since the FAO's data for cropland begins in 1961, cropland for 1960 is extrapolated using 1961 and 1962 data.

many of the major crops. For instance, corn (maize) and wheat yields increased by 391 and 208 percent, respectively (USDA 2000, 2005)

Figure 1.2 shows the contrast between trends in per-capita cropland, irrigation water use and irrigated land in the USA. In combination, figures 1.1 and 1.2 demonstrate that cropland per capita has declined at least since 1910, while aggregate cropland rose slightly until around 1930 and then fell very gradually. Although both aggregate and per capita levels of water use for irrigation have declined since around 1980, this followed substantial increases throughout most of the 20th century.

Overall, between 1910 and 2000 irrigation water use per capita and irrigated land per capita increased 15 and 55 percent, respectively. By comparison, over the same period, cropland per capita declined by 68 percent. Figures 1.1 and 1.2 show that between 1910 and 1950, US irrigation water use grew more rapidly than irri-

Figure 1.4 Global per capita cropland and irrigation water use 1900–2000

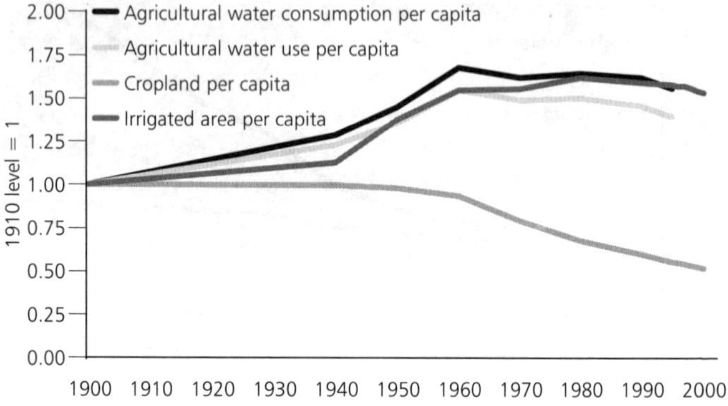

Sources: Shiklomanov (2000); McEvedy and Jones (1978); Goklany (1999a); FAO (2002).

gated land, but this trend was reversed in the 1950s. Currently, irrigated land seems to be increasing at a faster rate than irrigated water use.

Global trends: 1900 to 2000

Figure 1.3 shows global trends in aggregate land and water use and consumption by agriculture between 1900 and 2000. It suggests that they are on paths similar to that of the United States, except not as far along. While the amount of land devoted to crops seems to be levelling off (Goklany 2001a), agricultural water use and consumption, and the area devoted to irrigated land, continue to increase, although much less rapidly now than in the past.

Moreover, during this period, total water use and consumption of water relative to population growth have both increased much more than the amount of cropland. Between 1900 and 1995, the global population increased by 249 percent, cropland increased 95 percent, and agricultural water use increased 388 percent. Agricultural water consumption and irrigated land area increased even faster – by 446 and 453 percent, respectively.[3]

Figure 1.4 provides the same information, but on a per capita basis. It shows that cropland per capita has declined since around the 1930s. Between 1900 and 1995, it fell by 44 percent. By contrast, per capita agricultural water use and consumption both peaked around 1960. Although they have declined since then, per-capita withdrawals, per-capita water consumption due to agriculture and per-capita irrigated land were higher in 1995 than in 1900 (by 40, 56, and 58 percent, respectively).

Just as for the US, Figures 1.3 and 1.4 show that global agricultural water withdrawals and consumption grew more rapidly than irrigated land in the first four decades of the 20th Century, but since then this trend has reversed. Since 1980, irrigated land has increased at a faster rate than either agricultural water withdrawals or consumption. Between 1980 and 1995, irrigated land area increased 25 percent, while water withdrawals and consumption increased by 19 and 21 percent, respectively.

The environmental transition hypothesis

What accounts for the large differences in the trends for agricultural water and land use in both the US and worldwide? Why has agricultural water use increased much more rapidly than land use? Why did increases in the efficiency of cropland use precede those of agricultural water use?

The trends displayed in Figures 1.1 through 1.4 are consistent with the "environmental transition hypothesis" (Goklany 1998; 1999b). This is depicted graphically in Figure 1.5: the y-axis indicates the environmental impact (EI) on a society as measured by a particular indicator (such as air quality), while the x-axis represents time (which is a proxy for the state of technological development). EI first increases, then it goes through an "environmental transition" (ET) after which it declines – at least until society determines that it is

Figure 1.5 **The environmental transition**

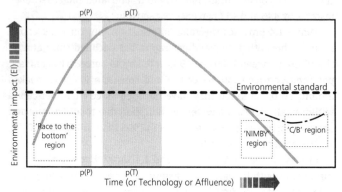

Note: p(P) = period of perception; p(T) = period of transition; NIMBY region = "not in my back yard" region (EI enters this region if benefits far exceed costs borne by beneficiaries); C/B region = EI enters this region if costs and benefits have to be more carefully balanced.
Source: Goklany (1999b).

'clean enough' (Goklany 1999b). Until that point, the trajectory for EI is shaped like an inverted-U.

For some indicators, such as sanitation or safe water, the transition has historically occurred early on in a country's developmental history (Goklany 1995). Currently-available trend data might therefore start at a point after the transition has occurred. That is, the trend data may only indicate the downward slope. For other indicators, because the problem has yet to be addressed successfully, a transition may not be evident, in which case the country may still be on the upward slope of the ET.

Historical trends for a variety of environmental indicators in many of the world's wealthier countries follow a path stylistically shown in Figure 1.5. These include various indicators related to air quality for traditional air pollutants, such as lead, sulfur dioxide, particulate matter, and carbon monoxide (Goklany 1999b), as well as indicators of water quality, such as dissolved oxygen levels, lead, and DDT (Goklany 1994; 1998; European Environment Agency 1998).

However, data from relatively poorer countries often shows that their pollution levels are currently increasing – that is, they are on the ascendant part of the environmental transition curve (Goklany 1994). So what accounts for environmental transitions?

Human beings are on a continual quest to improve their quality of life, which is determined by numerous social, economic, and environmental factors (Goklany 1995; 1998; 1999b). The weight given to each determinant changes constantly along with a society's precise circumstances and perceptions at any given point in time.

Over time, people develop new and better technologies with which to satisfy their needs and wants. The improved technologies typically increase productivity and lead to increases in wealth. In the early stages of economic and technological development, society places a higher priority upon satisfying basic needs, such as food, water, shelter, and heat, than on other concerns. This is true even if this higher priority entails tolerating some environmental deterioration.

Fundamentally, in the early phases of economic development people place a higher priority on affluence, because it is a means of achieving all their basic needs and also provides the means to reduce the most significant risks to public health and safety, such as malnutrition, infectious and parasitic diseases, and child and maternal mortality.

Also, in these early stages, society may be unaware of the potential risks posed by any particular technology. However, as society becomes wealthier, it tackles the more significant problems and possibly gains more knowledge. Hence, the specific environmental problems represented by EI automatically rise higher on its priority list (even if EI does not become worse).

In addition, because economic activity frequently increases EI, people perceive improvements to environmental quality as a more important determinant of the overall quality of life. This stage is represented in Figure 1.5 as the period of perception or p(P) (Goklany 1999b). Prior to p(P) one should not expect society to require, or private parties to volunteer, to reduce EI, although reductions may occur as a result of secular improvements in technology or other reasons (Goklany 1995;1996).

For example, in the US, the period of perception for sulphur dioxide probably did not begin earlier than October 1948, when an air pollution episode in Donora, Pennsylvania, was associated with 18 excess deaths in a population of 14,000. Nevertheless, for reasons explained below, indoor sulphur dioxide levels began to improve before the 1940s (Goklany 1999b).

From p(P) onward, a democratic society will often translate its desire for a cleaner (or improved) environment into laws, either because improvements are not forthcoming voluntarily or rapidly enough, or because of sheer symbolism. In general, this means that a relatively wealthier society will have more demanding laws, and will be more able to bear the costs of those laws.

At the same time, with increasing affluence and ongoing improvements in technology, society is better able to improve its environmental quality. Affluence also liberates the resources required to engage in research and development targeted towards cleaner technologies. Likewise, affluence makes possible the widespread deployment of new, or existing but unused, technologies (especially if their up-front costs are higher) by individuals in a society. Consequently, EI undergoes a period of transition. Ultimately, greater affluence and technological change should result in a decline in EI (Goklany 1995; 1999b).

The timing, height, and width of an environmental transition for a specific indicator is unlikely to be the same for all countries. In general – all else being equal – the ETs for latecomers to industrialisation should occur at lower levels of affluence because they can learn and adapt technologies from countries that have already gone through the ET. Indeed, this seems to be the case worldwide. By comparison with relatively wealthier countries, many of today's poorer countries have started to address environmental issues (such as safe water and sanitation, and lead, sulfur dioxide, and other air pollutants) at much lower levels of economic development. Sometimes, these countries are actually cleaner (or better off) than wealthier countries were at equivalent levels of economic development (Goklany 1995; 1999b; 2001a).

Other factors can also affect the timing of an environmental transition in a country, and the level of affluence at which it occurs. First, they depend on the precise indicator used to characterize environmental impact, and how closely it is tied to the perceived quality of life. This helps explain why in the US, for example, the transitions occurred earlier for indoor air pollution than for outdoor air quality, and for sulphur dioxide and particulate matter (pollutants most directly related to the killer air pollution episodes of the 1940s and 1950s) than for less powerful pollutants such as nitrogen oxides and ozone (Goklany 1999b). Similarly, we see that worldwide, countries address the lack of safe water and poor sanitation ahead of other forms of water pollution.

Second, the timing, height, and width of an environmental transition also depends upon government responsiveness to the perceived needs and desires of the general public. Thus democracies are more likely to see earlier transitions. In addition, the relative political power of the sectors which contribute environmental impacts can affect the timing, height and width of the transition, because that determines their success in affecting the stringency of laws directed at their contribution to EI.

Environmental transitions are also affected by the country's natural resource endowment. A country endowed with abundant natural gas and hydropower is less likely to burn coal while a country with large and easily accessible reserves of coal will be less eager to switch to cleaner fuels. Similarly, the period of perception for water use is more likely to be delayed for a country with plenty of water, for instance.

Finally, the extent of a transition will depend on the fiscal resources and human capital available and devoted to bringing about the environmental transition. The costlier or more difficult it is to bring about a transition, the more likely it will be delayed, the period of transition may be longer, and the height at which the transition occurs will be greater.

This helps to explain why relatively poorer countries are still on the upward slope for many pollutants even though wealthier countries have gone past their transitions. It also helps to explain why greenhouse gas emissions, for example, continue to increase globally (Goklany 1999b).

A superficially similar result is obtained by plotting environmental impacts against affluence (GDP per capita) using cross-country data for a range of pollutants. In this case, the inverted-U shaped curve has been called an "Environmental Kuznets Curve" (EKC) after the Nobel prize winning economist Simon Kuznets.

However there are significant differences between the ET and the EKC. In the former, the environmental indicator is plotted for one country (or region) at a time, and the x-axis represents time, which is a good proxy for both affluence and technological development, at least for the past two centuries (Goklany 1999a; 1999b). However, in the EKC, the data are generally plotted for a set of countries, and the x-axis represents only affluence.[4]

With respect to water resources, one measure of environmental impact could be the irrigation water withdrawn or consumed. Similarly, cropland use could serve as a measure of the human impact on the land (Goklany 1996). However, the amount of irrigated land does not fit neatly as a measure of environmental impact solely for either land or water, although it might help to explain some of the trends in land and water use.

US situation

Figures 1.1 and 1.2 show that for the US, aggregate as well as per capita irrigation water use have gone beyond their environmental transitions (or peaks). Aggregate cropland is close to, and perhaps also past the transition while cropland per capita is clearly past its environmental transition. These figures also indicate that although aggregate cropland has stayed more or less static for the 20th Century, the increase in the productivity of agricultural land use substantially exceeds the increase in water use productivity.

One possible reason as to why the decline in cropland per capita commenced earlier than agricultural water use per capita might be that cropland, in contrast to water, has mainly been privately owned. While there are several reasons why this has traditionally been the case (e.g., water supplies are uncertain and variable, not all its uses are rival, and water use can result in externalities),[5] private property rights over land provide the owner with powerful incentives to maximize long term productivity per unit of land (this is discussed further below). These incentives are weaker where private property rights are either absent, poorly delineated, or difficult to enforce – as has been the case with water in the US.

If US agricultural technology had been frozen at 1910 levels – i.e. if cropland per capita had stayed at 1910 levels – then to produce the same output as achieved in 2004, US farmers would have had to utilise 1,007 million acres rather than the 305 million acres that were actually harvested that year.[6] That's more than four times the total amount of land and habitat under special protection in the US in 1999 – including National Parks, National Wildlife Refuges, and National Wilderness Areas. Quite possibly, the increase in land productivity averted a potential catastrophe for US wildlife and perhaps even biodiversity more generally.

By contrast, water use per capita increased between 1910 and 2000, possibly because water use is more dependent on political muscle and machinations than on economics. Once access to water has been secured, in the absence of the ability to sell excess water or transfer it to other users for compensation, there are limited incentives to increase the efficiency of water used in agricultural activities.

However, even where *de facto* water "rights" are not fully transferable, there is an incentive to optimise water use within such constraints. One method to optimise water use is to improve irrigation efficiency which, in turn, would allow more land to be irrigated.

In the early part of the 20th Century, farmers and the agricultural sector had established a more-or-less free reign over water in the US. However, throughout the 20th Century, the demographic and economic power of the agricultural sector declined, while that of urban, suburban, and environmental interests – interests with broad overlap in membership – increased:

- In 1899–1903, agriculture contributed 18.9 percent of US national income and employed 36.9 percent of the working population.
- By 1948–1953, agriculture contributed 7.2 percent of national income and employed 10.6 percent.
- In 1970, agriculture accounted for 3.1 percent of national income and employed 4.3 percent of the working population. (US Bureau of the Census 1975).

At the same time, the percentage of the population living in rural areas declined from 60 percent in 1900, to 41 percent in 1950, and 26 percent in 1970 (US Bureau of the Census, 1975). Also by 1970, the demand for water and the costs of tapping new sources of water had increased for all sectors.

Thus, politics and economics came together in a way that enabled the urban-suburban environmental groups frequently to challenge agriculture's claims to water. Though not all of these challenges were fully successful, by the 1980s they had served to reduce the amount of water diverted to agriculture, as well as irrigation water use per capita (Solley et al. 1998; Postel 1999). The agricultural sector responded to increased water scarcity by increasing the efficiency of irrigation and expanding the amount of land under irrigation.

This helps explain both the decline in the amount of irrigation water applied per acre of land – from about 2.5 acre-feet in 1980 to 2.1 acre-feet in 1995 (Solley et al. 1998) – and the rapid increase in irrigated land during this period, even as irrigation water use declined.

Global situation

Figure 1.3 shows that globally, aggregate cropland seems to be levelling off – i.e. approaching an environmental transition (Goklany 2001a). But there is still an increase in aggregate water use and consumption as well as in irrigated land use (albeit less rapid than previously). Moreover, except for cropland, they have all increased at a faster rate than population.

On a per-capita basis, however, cropland and irrigation water use and consumption have all passed their environmental transitions. But these levels have not yet dropped off as much as the levels for the US.

Despite the pressures which agriculture has brought to bear on global biological resources, those pressures could have been much worse if global agricultural productivity, and therefore yields, had been frozen at a certain point in time (for instance, at 1961 levels). This would be equivalent to freezing technology, and its penetration, at 1961 levels. In that case, in order to produce as much food as was actually produced in 1998, agricultural land area would have had to more-than double to at least 26.3 billion acres, compared to the actual 1998 level of 12.2 billion acres (Goklany 2001b).

Thus, agricultural land area would have needed to increase from its current 38 percent to 82 percent of global land area (FAO 2001; Goklany 2001b). Cropland would also have had to more than double, from 3.7 to 7.9 billion acres. In effect, an additional area equivalent to the size of South America-minus-Chile would have been ploughed-under to achieve the 1998 level of food production. Thus, by increasing productivity in land, we forestalled further increases in threats to terrestrial habitats and biodiversity.

However, these improvements were not matched by similar increases in efficiencies of irrigation water use. Not surprisingly, some analysts now believe that the major resource constraint for being able to satisfy future global demand for food is likely to be water rather than land, as Malthus and others had traditionally believed (FAO 1996; Postel et al. 1996; Pimentel et al. 1997; Postel 2000).

Property rights – and their absence

Whether considering the US or the global situation, water use efficiency has lagged behind improvements in cropland efficiency for similar reasons. Namely, in most areas of the world farmers possess some property rights to their land but often not to water; nor is water generally treated as an economic commodity.

In fact, the tremendous increase in irrigation in the US (Gleick 1998) and worldwide during the past few centuries (L'Vovich et al. 1990; Goklany 1998) could be viewed – at least in part – as the substitution of often-subsidized water for land. This provides evidence for Terry Anderson's statement that when water is cheaper than dirt, it will be treated that way (Anderson 1995).

Property rights include long-term tenure to land, the right to trade, and the right to profits from selling products and improving productivity (Goklany and Sprague 1991; IPCC 1991; Taylor 1997). Farmers would not invest – i.e. risk – their time, money, and effort to increase productivity and efficiency without such rights, especially the right to retain profits from such investments. Property rights therefore provide an incentive for the farmer to engage in long term sustainable practices.

A good example of the beneficial effects of property rights comes from the improvements in agricultural productivity in China in the early 1980s, and the subsequent slowdown in improving yields in that country (Prosterman et al. 1996). In the early 1980s, Chinese farmers were given rights to a portion of their produce. The annual rate of increase in agricultural productivity soared. However, it declined again when it became clear that these rights were not equivalent to long term tenure, and farmers held back further investments in "their" plots.

Property rights are one important aspect of "economic freedom." Indeed, Norton (1998) has argued convincingly that economic freedom serves as an aggregate measure for the deference given by a country to property rights, since it includes components for the security given to property rights under law, as well as components which would diminish those rights indirectly through inflation or through limitations on the freedom to trade or exchange.

Not surprisingly, Gwartney et al. (1998) find that increases in cereal yields are proportional to the degree of economic freedom in countries. Norton also finds that deforestation rates decline when property rights are increased. These two sets of results – increased yields and lowered deforestation – support the notion that stronger property rights result in higher agricultural productivity, which leads to greater land conservation (Goklany and Sprague 1991; IPCC 1991; Goklany 1998; 1999a).

On the other hand, the absence of property rights for water simply encourages waste and reduces incentives to adopt existing or develop new technologies to enable conservation, re-use or recycling. To make matters worse, most societies subsidise water – particularly in agriculture – on the basis that water is vital for humanity (Anderson 1995; Pimentel et al. 1997). Such subsidies perversely reduce the incentive to utilise water more efficiently, or to conserve – and predictably, water conservation technologies remain under-utilized and under-developed.

Yet another perverse consequence of water subsidies is that in many urban areas in the developing world, the poor pay more for water than do the middle and upper classes that are connected to subsidized municipal water systems (Serageldin 1995). Ironically, many in these subsidized groups are happy enough to pay larger sums for bottled or canned soft drinks even when they are not needed to quench their thirst.

If institutions and policies are modified so that private entities are able to hold well-defined and transferable rights to water, this would form the basis for water markets. Water trading would occur, the price of water in specific contexts would reflect demand and scarcity, and 'virtual' water would be freed up. It might then be possible to replicate for water the almost universal historical experience with land, in which use has progressively has become more efficient.

The success of such policies and approaches has been demonstrated in cultures as diverse as the US, Chile, Jordan, India, Pakistan, and Indonesia (Anderson 1995; Rosegrant et al., 1995; Serageldin, 1995; Easter et al., 1998).

For example, in Chile, efficiency of water use was increased through water trading. Between 1976 and 1992, it improved from 22 to 26 percent, effectively expanding irrigated area by a similar amount (Rosegrant et al. 1995; see also Southgate and Figueroa, this volume, Chapter 3). The experience in India and Pakistan shows that gains in efficiency can be obtained even where markets are based on informal and imperfect property rights (Saleth 1998; Meinzen-Dicks 1998).

The environmental benefits of property rights

The environmental benefits of well defined, readily enforceable and easily transferable property rights are evident in other arenas besides water. For example, with respect to air pollution, the initial major improvements in air quality were experienced in the first half of the 20th Century when households and businesses began to switch from coal- and wood-burning stoves and fireplaces to oil and gas, while others adopted more efficient combustion equipment and practices (Goklany 1996; 1999b).

By and large, homeowners and businesses undertook these measures voluntarily because they were cleaning up their own private property. They were confident that investments in more efficient, cleaner fuel delivery would result in direct benefits by reducing smoke, dust, and grit to themselves, their families, and, in the case of businesses, for their employees and customers. No less important was the fact that the development of new, more efficient technology reduced their fuel costs. Thus, by virtue of the institution of property rights, they had an economic as well as an environmental incentive to clean up and to use resources more efficiently.

The ability of property owners to capture the economic benefits associated with greater efficiency also provided much of the impetus behind the secular improvements in technology which helped to reduce emissions per unit of GDP for sulfur dioxide (SO_2), volatile organic compounds (VOCs), and nitrogen oxide (NO_x) pollutants. These improvements occurred long before any of those substances were generally recognized to be environmental problems or, for that matter, before the US federal government became involved in air pollution control (Goklany 1999b).

SO_2 was not perceived to be a public health problem until after several 'killer smog' episodes – including well-known events in Donora, Pennsylvania in 1948, and London in December, 1952. Yet SO2 emissions per unit of GDP have been declining since the early 1920s in the US while London's outdoor sulphur dioxide concentrations peaked in the late 1890s (Elsom 1995; Brimblecombe 2004).

Similarly, emissions of VOCs and NOx per unit of GNP have been dropping across the US since the 1930s – decades before these substances were either implicated (in the 1950s) as being responsible for the formation of photochemical smog, or recognized (in the late 1960s and early 1970s) to be nationwide air quality problems (Goklany 1999b). Likewise, CO_2 emissions per unit of GDP in the U.S. have been declining at the rate of 1.3 percent per year for the past century and a half – long before global warming registered on the public consciousness in the late 1980s.

Summary and conclusions

Currently, agricultural land use and water use constitute some of the primary pressures on terrestrial and freshwater habitats, ecosystems and biodiversity, both in the USA and globally. These pressures would be much worse but for the increased productivity of land in agriculture during the 20th Century.

This leads to one of the most interesting conundrums concerning natural resource use. Although, decade-upon-decade, agricultural productivity per unit of land in the US and worldwide has increased spectacularly, these increases have not been matched by parallel improvements in productivity per unit of water.

It seems that water has been systematically substituted for land in order to boost productivity. As a result, the availability of water in the 21st Century could become the major impediment to resolving the dilemma inherent in satisfying global food and fibre needs while conserving habitat and maintaining biodiversity.

The differences in the trends for agricultural land and water use can, partly at least, be explained by the almost universal difference in the institutional arrangements for the use and management of these two critical natural resources. The widespread institution of clearly defined and readily enforceable and transferable property rights over agricultural land has provided the incentives and capital that have enabled enhanced efficiencies.

If similarly well-defined and readily enforceable and transferable property rights were to be instituted for water, it seems likely that entrepreneurs operating within the market would likewise increase the efficiency of use of water for agriculture – thereby helping solve the food/biodiversity dilemma by enabling both to be provided in ample amounts.

Notes

1. Farms account for about half of the land area outside of Alaska (which has very little agricultural potential), about one-third of which is harvested each year.
2. However, a word of caution is needed regarding these data. Figure 1.1 is interpolated from data provided in the periodic Censuses of Agriculture gathered until 1997 by the Department of Commerce (US Bureau of the Census 1975; US Department of Agriculture 2001b). The Census of Agriculture – which was gathered by the US Department of Agriculture for the first time in 1997 – shows a very rapid increase since the early 1990s in the amount of irrigated land. However, another data set collected by the US Geological Survey (Solley et al. 1998) shows a much smaller increase. A comparison of estimates made by the Census of Agriculture (interpolated) and those of the USGS show that the latter's figures are consistently higher by 10 to 25 percent.

3. The amount of irrigated land tracks relatively closely with agricultural water consumption because the two sets of data are related and come from the same source (i.e., Shiklomanov 2000).
4. In fact, I have elsewhere demonstrated that a set of single-country inverted-U-shaped ETs does not necessarily result in an inverted-U-shaped cross-country EI verses affluence curve; instead, it could be N- or even U-shaped (Goklany 1999b).
5. See for example Livingston (1998).
6. This calculation is based on three relatively optimistic assumptions: First, sufficient new cropland would be available, but this is unlikely since the total amount of potential cropland in the US is estimated to be only 647 million acres (Goklany 2001a, based on NRCS 2001). Second, the additional cropland would be just as productive as existing cropland. Third, the productivity of existing cropland would be maintained without any new technologies.

References

Anderson, Terry L. (1995). Water, Water Everywhere But Not a Drop to Sell. In: *The State of Humanity*, ed. Julian Simon, E. Calvin Beisner and John Phelps, Cambridge, MA: Blackwell, 425–433.

Brimblecombe, Peter (2004). Perceptions and Effects of Late Victorian Air Pollution. In: *Smoke and Mirrors: The Politics and Culture of Air Pollution*, ed. E. Melanie DuPuis, New York: New York University Press, 15–26.

Easter, K. William, Mark W. Rosegrant, and Ariel Dinar (1998). *Markets for Water: Potential and Performance*. Norwell, MA: Kluwer Academic Publishers, 298 pages.

Elsom, Derek (1995). In: *The State of Humanity*, ed. Julian Simon, E. Calvin Beisner and John Phelps, Cambridge, MA: Blackwell, 476–490.

European Environment Agency (1998). *Europe's Environment: The Second Assessment*. Oxford, UK: Elsevier Science.

Food and Agricultural Organisation (FAO)(1996). *Water and Food Security*. World Food Summit Fact Sheets. Rome, Italy: FAO.

——— (2002). *FAOSTAT Database, 2002*. Available at: http://apps.fao.org/. Visited 6 December 2002.

——— (2001). *FAOSTAT Database, 2001*. Available at: http://apps.fao.org/. Visited 3 October 2001.

Gleick, Peter H. (1998). *The World's Water: The Biennial Report on Freshwater Resources*. Washington, DC: Island Press.

Goklany, Indur M. (1994). *Air and Inland Surface Water Quality: Long Term Trends and Relationship to Affluence.* Washington, DC: Office of Program Analysis, US Department of the Interior.

—— (1995). Richer is Cleaner: Long-Term Trends in Global Air Quality. In: *The True State of the Planet*, ed., Ronald Bailey, New York, NY: Free Press, 339–377.

—— (1996). Factors Affecting Environmental Impacts: The Effects of Technology on Long Term Trends in Cropland, Air Pollution and Water-related Diseases." *Ambio* 25: 497–503.

—— (1998). Saving Habitat and Conserving Biodiversity on a Crowded Planet. *BioScience* 48: 941–53.

—— (1999a). Meeting Global Food Needs: The Environmental Trade-Offs Between Increasing Land Conversion and Land Productivity. *Technology* 6:107–30.

—— (1999b). *Clearing the Air: The Real Story of the War on Air Pollution.* Washington, DC: Cato Institute.

—— (2001a). *Extending the Limits to Growth.* Unpublished manuscript.

—— (2001b). Agricultural Technology and the Precautionary Principle. Political Economy Research Forum, November 29–December 2, 2001, Bozeman, MT: Political Economy Research Forum.

Goklany, I. M., and M. W. Sprague (1991). *An Alternative Approach to Sustainable Development. Conserving Forests, Habitat and Biological Diversity by Increasing the Efficiency and Productivity of Land Utilization.* Washington, DC: Office of Program Analysis, US Dept. of the Interior.

Gollehon, N., W. Quinby, and M. Aillery (2003). Agricultural Resources and Indicators: Water Use and Pricing in Agriculture. In: R. Heimlich, ed., *Agricultural Resources and Environmental Indicators 2003*. Available at http://www.ers.usda.gov/publications/arei/ah722/, visited August 19, 2005.

Gwartney, James, Randall Holcombe, and Robert Lawson (1998). The Scope of Government and the Wealth of Nations. *Cato Journal* 18: 163–90.

Hutson, Susan S., Nancy L. Barber, Joan F. Kenny, Kristin S. Linsey, Deborah S. Lumia, and Molly A. Maupin, *Estimated Use of Water in the United States in 2000*, US Geological Survey (USGS) Circular 1268 (released March 2004, revised April 2004, May 2004, February 2005). Available online at: http://pubs.usgs.gov/circ/2004/circ1268/, visited January 31, 2006.

Intergovernmental Panel on Climate Change (IPCC) (1991). Resource Use and Management. Chapter 6. In *Climate Change: The IPCC Response Strategies*. Washington, DC: Island Press, 163–202.

IUCN (1999). *The Freshwater Biodiversity Crisis*. World Conservation 2/99. Available at http://www.iucn.org/bookstore/bulletin/1999/wc2/content/freshwaterbio.pdf. Visited 7 November 2001.

—— (2000). *Confirming the Global Extinction Crisis*. IUCN Press Release, 28 September 2000. Available at http://www.iucn.org/redlist/2000/news.html. Visited 7 November 2001.

Livingston, Marie Leigh (1998). Institutional Requisites for Efficient Water Markets. In: *Markets for Water: Potential and Performance*, eds. K. W. Easter et al. Norwell, MA: Kluwer Academic Publishers, 19–33.

L'Vovich, Mark I., Gilbert F. White, A. V. Belyaev, Janasz Kindler, N. I. Koronkevic, Terence R. Lee, and G. V. Voropaev (1990). Use and Transformation of Terrestrial Water Systems. In: *The Earth as Transformed by Human Action*, eds. B. L. Turner II et al. New York: Cambridge University Press, 235–52.

Maddison, Angus (2001). *The World Economy: A Millennial Perspective*. Paris: Organization of Economic Cooperation and Development.

McEvedy, C, and R. Jones (1978). *Atlas of World Population History*. New York: Penguin.

Meinzen-Dicks, Ruth S. (1998). Groundwater Markets in Pakistan: Institutional Development and Productivity Impacts. In: *Markets for Water: Potential and Performance*, eds. K. W. Easter et al. Norwell, MA: Kluwer Academic Publishers, 207–222.

NRCS (Natural Resources Conservation Service) (2001). *Summary Report: 1997 National Resources Inventory, Revised December 2000*. USDA-NCRS. Available at http://www.nhq.nrcs.usda.gov/NRI/1997/summary_report/original/contents.html. Visited 11 October 2001.

Norton, Seth W. (1998). Poverty, Property Rights, and Human Well-Being: A Cross-National Study. *Cato Journal* 18 (2): 233–245.

Pimentel, D., J. Houser, E. Preiss, O. White, H. Fang, L. Mesnick, T. Barsky, S. Tariche, J. Schreck, and S. Alpert (1997). Water Resources: Agriculture, the Environment, and Society. *BioScience* 47(2): 97–106.

Postel, S., G. C. Daily, and P. R. Ehrlich (1996). Human Appropriation of Renewable Fresh Water. *Science* 271: 785–88.

Postel, Sandra. (1999). *Pillar of Sand: Can the Irrigation Miracle Last?* New York: W. W. Norton.

Prosterman, R. L., T. Hanstad, and L. Ping (1996). Can China Feed Itself? *Scientific American* 275(5): 90–96.

Rosegrant, M.W., R. G. Schleyer, and S. N. Yadav (1995). Water Policy for Efficient Agricultural Diversification: Market-based Approaches. *Food Policy* 20: 203–223.

Saleth, R. Maria (1998). Water Markets in India: Economic and Institutional Aspects. In: *Markets for Water: Potential and Performance*, eds. K. W. Easter et al. Norwell, MA: Kluwer Academic Publishers, 187–205.

Serageldin, Ismail (1995). *Toward Sustainable Management of Water Resources*. Washington (DC): World Bank.

Shiklomanov, Igor. A. (2000). Appraisal and Assessment of World Water Resources. *Water International* 25: 11–32.

Solley, Wayne B., Robert R. Pierce, and Howard A. Perlman (1998). *Estimated Use of Water in the United States in 1995*. US Geological Survey Circular 1200. Denver, CO: US Geological Survey.

Taylor, C. (1997). The Challenge of African Elephant Conservation. *Conservation Issues* 4 (2): 1, 3–11.

US Bureau of the Census (1975). *Historical Statistics of the United States: Colonial Times to 1970*. Washington, DC: Department of Commerce.

US Bureau of the Census (2006). *Statistical Abstract of the United States 2006*. Washington, DC.

US Department of Agriculture (2000). *Crops: US Estimate Track (Series) Record, 1866–Current*. National Agricultural Statistics Service (NASS), stock no. 96120. Available at: http://www.ers.usda.gov/prodsrvs/dp-fc.htm. Visited: 27 July 2000.

US Department of Agriculture (2001a). *Historical Track Records*. National Agricultural Statistics Service (NASS). Available at: http://www.usda.gov/nass/pubs/backrec/trackrec2001.pdf. Visited: 20 December 2001.

US Department of Agriculture (2001b). *Agricultural Statistics 2001*.

US Department of Agriculture (2005a). *Agricultural Statistics 2005*.
US Department of Agriculture (2005b). *USDA Data on Major Land Uses*, available at http://www.ers.usda.gov/data/MajorLandUses/spreadsheets/c1910_00.xls. Visited: 20 December 2005

Wilson, Edward O. (1992). *The Diversity of Life*. Cambridge (MA): Belknap.

2 Incentives matter: The case for market valuation of water

Andrew P. Morriss[1]

Is it possible to value water? How are we to do so? Indeed, how does anyone value anything? There are three broad categories of answers to these questions.

One relies on the notion of an intrinsic value of a good, which suggests that a particular good has a definitive value which can be determined in some way. Theological speculations on the 'just price' of a good and other speculations on 'innate values' of various things often fall into this category (Woods Jr. 2005). A second category denies that a particular good is capable of valuation. Human lives, for example, are often classified as being incapable of valuation (Dietz & Vollebergh 1999, 344). The third category depends on observations about human behavior, in particular, observations about the actual choices made by real people with their scarce resources.

Markets fall into this third category. I contend that markets are connected with real choices made by real people, which lead to real consequences. These attributes enable markets to value resources in a superior manner compared to alternative arrangements.

Indeed, this chapter argues that markets provide the *only* way to value resources, including water, in a manner which does not provoke conflicts among competing users.

There are four important reasons why it is appropriate to value water (and other resources) with market processes.

First, markets are a low cost means to provide important signals to people about the value of various uses of water. These signals help water flow to the uses where it produces the largest net benefit for water users. Second, markets allow the uses of water to vary with changes in knowledge and demand. Because they provide a dynamic – rather than a static – valuation, markets adapt to constantly changing circumstances. Third, markets encourage the production of new knowledge about water: new uses and new ways to think about these uses, new conservation methods, new delivery methods, etc. Thus, markets encourage investment in meeting human needs. Fourth, markets do not require large scale agreement among their participants on overall ends, allowing a diversity of individual ends to coexist peacefully.

In large part, the argument about markets for water is about governance of resources more generally. Markets excel in generating positive-sum transactions that benefit all parties – precisely the situation in which we should all aspire to be.

MARKETS AS LOW COST SIGNAL MECHANISMS

Transactions costs affect the ways that people organize their activities. They are real, concrete costs – such as the costs of making and enforcing contracts, and the cost of creating and enforcing property rights – which affect every day decision-making. Relative transactions costs determine whether or not people choose to conduct a transaction within a firm or in the marketplace, and whether or not it is possible to contract around an inefficient legal assignment of rights.

The cost of transacting is thus a critical component of any method which assigns value. Information costs are an important part of transactions costs (Morriss et al. 2002, 337). Transactions themselves also provide important information to others about the parties' valuation of resources, giving them a means to assess the value of their own assets and the costs of possible projects they are considering undertaking.

A market price is a remarkably compact source of fairly dense information. The market price of a commodity summarizes in a concise way (the price itself plus the terms at which the price is available) the current valuation of the good being priced (Hayek 1945; Wills 1997, 31). That valuation involves the various uses of the good in question plus the alternative uses of the resources used to make the good. The fact that markets can summarize information about myriad alternative uses of water into an easy-to-comprehend number – a price – is an all-important virtue.

Low cost transmission of information, particularly of complex information, is valuable for several reasons.

1. Less costly information means more transactions

It is a transaction cost to learn about the relative scarcity of a resource. When transactions costs fall, it means there are fewer barriers to transactions. When more transactions occur, it means that more voluntary trades occur; and more voluntary trades produce more wealth, for trade increases wealth. Increasing wealth is not only desirable in its own right but also often increases demand for improved environmental quality.

Institutions that reduce the cost of information increase the volume of transactions by lowering the cost of transacting. Today's paradigmatic example is the significant reduction in information costs produced by the dramatic drop in telecommunications costs. In turn, this was brought about by a combination of the significant steps toward deregulation of telecommunications (Hahn and Hird 1991, 233, 250), and the spread of the internet as a means of communicating price information.

All else equal, more voluntary transactions increase net social welfare because individuals do not engage in voluntary transactions unless they increase the welfare of the transacting parties. eBay is a well-known contemporary example of how increasing the volume of transactions increases welfare (Grimmelman 2005, 1749). Using eBay, individuals have found a market for countless items which are valued more highly by their buyers than by their previous owners. By broadening the potential market for these items, eBay has increased the number of transactions which are possible.

eBay does this primarily through three market-enhancements which lower transactions costs and thus increase the number of transactions. First, it reduces the search costs of finding items that buyers want. Second, it uses a feedback mechanism to provide a means to determine the quality of the buyer and seller. Third, through its PayPal subsidiary, eBay lowers the transactions costs of making payments. These features of the eBay market increase net welfare as goods move to users who value them more highly.

Water is no different from Beanie Babies™ in this regard. In the absence of a market for water, one of the problems which hinder mutually beneficial trades between users (for instance, two figurative parties called Smith and Jones) is that they lack the information needed to know that the transaction is possible, and that the transaction will produce an increase in their own welfare. Smith and Jones, in addition to people outside this transaction, have no way to know who values the water more. Even if we ask them to tell us, and then use some kind of administrative mechanism to award it to the most highly valued use, we don't know whether Smith and Jones have truthfully revealed their preferences. (Smith and Jones cannot be trusted to honestly reveal their preferences unless dishonesty is costly.) Further, because most studies show that demand for water is price sensitive, Smith's and Jones' preferences are themselves dependent on the cost of water. In the absence of a market, Smith and Jones are *unable* to compare their values for the water in question (Woods 2005, 21).

Smith wants to water her lawn; Jones wants a drink: How do we know which use is more valuable, or how a wider set of individuals would prioritize those uses? Without a market to allocate the water between Smith and Jones, we must use some alternative mechanism. Without prices, administrative allocations often rely on crude proxies for use-value. Thus residential users often pay the same price for the water they drink and the water used for their lawns, even though watering a residential lawn would likely be a lower valued use.

Sorting through individual valuations is also a costly exercise for our token decision-maker, who must gather, weigh and analyze information and then consider the truthfulness of responses. In addition, the decision-maker may be tempted to weight his decision in favor of personal considerations: Is Smith a family relative? Does the decision-maker hope to work for Jones after leaving his position of authority? Did Smith or Jones vote for the decision-maker in the last election?

These considerations – which are wholly unrelated to the value of water – often play a role in determining administrative allocations of resources. It is also costly to guard against consideration of such factors – this entails a variety of steps to verify legitimacy, all of which ultimately reduce the efficiency of the decision-making process.

However, if Smith is willing to sell to Jones, then we know that Smith values the water less than she values the alternative resources she can purchase with the money which Jones gives her in exchange for the water. Likewise, if Jones is willing to pay a price at which Smith is willing to sell, we know that Jones values the water more than the resources he must give Smith to secure it. We can make this conclusion with confidence, because Smith and Jones have proven it to be true by their actions.

In contrast to the alternatives, markets provide information on relative valuations at a low cost. Alternative arrangements require that resources be expended to learn, evaluate and compare people's preferences. These additional costs reduce the number of transactions, which reduces net welfare.[2]

2. Low cost transmission of information reduces uncertainty

When the price of a good falls, people consume more of it. In this respect, information is no different from other goods. Because markets are a less costly way to convey information than are alternative means of ordering the world, they enable people to utilize more information in their transactions. This is an effective way to reduce uncertainty: with more information rather than less, people can more accurately predict the outcome of their activities and transactions, and bear less risk while transacting.

Risk-bearing is generally a costly activity (which is why we pay insurance companies to reduce our risk of loss). To the extent that people are risk averse (Wills 1997, 233) – and we have reason to believe that most people are risk averse in most situations – reducing risk increases human welfare (Shavell 1987). More information and less risk means that people can be more certain of achieving their goals: they can devote fewer resources to hedging against uncertainties, which means they have more resources available to spend in alternative uses (e.g. devote to accomplishing their goals).

Why are markets a less costly way to convey information? First, they convey information in a compact form, reducing a wide variety of information to a single number– a price – whose movements and value are readily understood by market participants. If the price of a good rises – or falls – people receive a clear signal that the resource is becoming more – or less – valuable.

Moreover, complex financial instruments offer even more sophisticated signals. For example, in a futures market for a given commodity, movements in the price of that good today can be compared to price movements in the future (Anderson and Snyder 1997, 192; hereinafter Anderson and Snyder). As the spread between the future price and the current price changes, markets convey information about the expectations of market participants about the future.

There is good reason to believe that collective expectations about the future compiled in market prices are far more accurate than individual expectations. Because prices are based on a wider information set, markets are inexpensive and more accurate alternatives compared to speculation by individuals about the availability of a resource in the future (Surowiecki 2004).

Second, the very existence of the market itself provides an incentive to lower transactions costs. The existence of such costs provides entrepreneurs with incentives to gain market advantages by reducing those costs. Further, differences in prices across markets give entrepreneurs an incentive to engage in arbitrage, which ultimately eliminates the differences in prices (except for differences due to the cost of the arbitrage itself).

Consider financial markets, which are among the most efficient markets known (Macey 1994, 927). In a remarkably short period of time, market prices on major stock markets fully reflect new information. The result is that investors can rely on financial market prices to provide accurate signals about the value of goods and services, and these signals can be used to construct strategies which reduce risk.

Among the most prominent examples of this phenomenon is Southwest Airlines' successful price hedging strategy for jet fuel. Southwest buys financial instruments which reduce its vulnerability to increases in fuel costs: these instruments pay off if the price of jet fuel increases, which offsets the higher price the airline must pay for jet fuel (*Wall Street Journal* 2005).

How might such an arrangement affect water markets? One issue for many water users is whether water will be available in the future to meet projected needs. Farmers, for example, benefit from knowing whether or not they will have access to water to irrigate their crops before they make decisions about planting their fields. Recreational water users, and the businesses that profit from selling them ancillary services, benefit from knowing whether sufficient water will be available for a time period in which they wish to plan recreation.

Predicting the future availability of water requires weather prediction, but the natural sciences have not yet developed highly accurate methods to create long range predictions. In the face of such uncertainty, water markets would utilize the same mechanisms currently used for some commodity crops to enhance availability of water in the future (Kreitner 2000, 1102; see also discussion of spot markets for agricultural water in arid regions of Chile, Southgate and Figueroa, this volume, Chapter 3). Not only would water users be able to reduce their risk by using water futures, but also the predictive accuracy of a large number of water users would likely be greater than the non-market predictions of scientists alone (Surowiecki 2004).

Finally, information tells us about the performance of an institution. Information on the performance of private water firms is less costly to acquire (because it is available more readily) than similar information for their non-market water counterparts (i.e. the state sector). Thus, it is far less costly to monitor the behavior of private firms, and this has clear benefits for their customers.

Not surprisingly, many state sector water institutions suffer from administration problems because they cannot effectively monitor their performance (Segerfeldt 2005, 21–22; hereinafter Segerfeldt; see also Robinson, this volume, Chapter 8). Indeed, public sector water institutions routinely fail at monitoring water use, one of the most basic information-related tasks. As a result, water is siphoned off and unaccounted for. This failure is not simply a failure of technology or competence, it is a failure caused by improper incentives (ibid. 23; see also Bhandari and Khare, this volume, Chapter 4).

3. Markets reduce the costs of information, which allows more complex transactions to occur

When information is less costly, transactions become more complex and more numerous. This enables individuals to accomplish their goals more precisely, which reduces transactions costs, reduces uncertainty and increases net welfare.

An important component of the cost of complexity is the cost of information. If individuals and firms only know one aspect of a commodity (e.g. quantity), then their transactions are limited to those which involve different amounts of the commodity. If, however, more aspects – such as the date and time of delivery – are known, then individuals and firms can vary their prices according to these numerous criteria.

When transactions become more complex, this enables individuals and firms to become more focused on the specific goals they seek to accomplish. In short, "[w]hen scarcity drives up the price of a resource, users of the resource are motivated to find alternative sources of supply, new technologies, or substitute resources" (Anderson and Snyder 18).

Water has a wide range of attributes which affect users in different ways. Consider the flow of water in a river: A unit of water in the river during a wet season has a different marginal value to the ecosystem than a unit during a dry season. The capacity of water systems can present challenges similar to those in electricity delivery, as 'peak' water costs more than 'off-peak' water as peak power costs more than off-peak power because of the greater marginal cost involved in operating at the system's limits (ibid.).

Water quality can also vary in a wide range of dimensions. Water which is allowed to stay in a river has effects on downstream users which are different than water which is removed from a river for use – even if the water is ultimately returned downstream. Water moved from one watershed to another by a user has different impacts than water which is returned to its original watershed.

It is not difficult to imagine a series of complex transactions involving water which could take place if such attributes were capable of being incorporated into markets. For example, recreational fishermen, conservation groups who are concerned with preserving an ecosystem, and tourism interests could combine to purchase rights to maintain minimum stream flows from agricultural users (ibid.124–130, 200–201). The agricultural users could, in turn, buy futures contracts that allowed them to hedge against low water levels that prevented abstraction of water for irrigation due to the stream flow contracts. The resulting transaction would be far more complex than a simple transfer of the right to a particular quantity of water. Such contracts could offer significant gains for all parties.

We have some historical evidence to support the idea that markets enable more complex transactions concerning water. During the nineteenth century, a large number of mutual irrigation companies were formed in the American West (ibid. 39–41). Economists Terry Anderson and Pamela Snyder write:

> *By using members' assets as collateral, mutual [irrigation companies] could enter capital markets to obtain the investment funds necessary to develop irrigation projects. The transferability of stocks ensured that water could be moved to higher valued alternatives, further ensuring the success of the operation. These*

features, combined with the security of rights provided by the doctrine of appropriation, stimulated an effective marketplace.

In more recent times, entrepreneurs have discovered ways to reduce waste water treatment costs through markets and complex transactions which involved multiple parties. For instance, an innovative treatment program enabled point- and non-point producers of waste to contract with each other in the Tar-Pimlico River Basin in Minnesota (Morriss et al. 2002, 345–46). These contracts saved millions of dollars in treatment costs. Others have found ways to extend water treatment systems without burdening municipal governments (Rapier 2005). By allowing more complex transactions, markets improve economic and environmental well-being.

Markets as dynamic signal mechanisms

One of the most important aspects of markets is their dynamic nature. Market prices quickly respond to constantly changing information. The change in prices sends signals to participants about the impact of events on the goods and services sold in markets. As discussed above, an individual can learn a great deal from market prices, and futures prices, about the likely course of events – even if that individual does not understand the underlying information which drives the changes in market prices (Thomsen 1994, 170). "The entrepreneur who discovers a higher valued use for water, for example, stands to gain from transferring his water to the higher valued use" (Anderson and Snyder 23).

Thus, markets are institutions which encourage individuals to adapt to constantly changing circumstances (Stroup 2000, 489).[3] Higher prices encourage new firms to enter an industry, expanding supply and often resulting in lower costs and higher quality for the consumer (Anderson and Snyder 11, 12–13). Because they tend to operate without such signals, public water companies often underprovide services to consumers (Segerfeldt 23).

The critical difference is that markets require individuals to face the 'opportunity costs' of their actions (Anderson and Snyder 23). An opportunity cost is the cost of what a resource owner foregoes by not taking an action.

For example, suppose 100 acre-feet of water can be used to irrigate a cotton field and produce a net benefit to the water owner of $2/acre-foot. The same amount of water could be used to irrigate a 50 percent less water-intensive crop, and the surplus could be sold for $3/acre-foot. In this case, the cotton farmer's opportunity cost of not changing to the less-water intensive crop is $50. If net total value of water sales and crop sales from the alternative crop exceeds the net total value of the cotton crop, and the cotton farmer does not know about the possibility, an entrepreneur will be able to bid more for the farmer's land than the farmer currently earns, by taking into account the knowledge of the potential for shifting crops and selling the surplus water (Segerfeldt 31; Anderson and Snyder 192–193).

Recognizing opportunity costs requires transferable property rights, for a water rights owner who cannot transfer her rights will never be able to consider the full opportunity costs of her actions: other parties which have alternative uses for the water rights will have no incentive to approach her with a proposed transfer (Anderson and Snyder 24).

Markets are also dynamic signal mechanisms because they rapidly incorporate new knowledge. Entrepreneurs quickly utilize new information when it becomes available, and they spread it to others through their own behavior.

For instance, if new data reveals that soybean harvests in China are likely to be poorer than expected, those who possess that information will seek to purchase soybean futures, knowing that (other things equal) the market price for soybeans will increase as the Chinese harvest data become more widely known. The rising price of soybean futures will provide information even to those who are completely ignorant of Chinese soybean production estimates, enabling them to learn and benefit from the information uncovered by the specialists.

In contrast, administrative allocations – which either forbid transfers outright or which depend on seeking regulatory approval – interfere with this dynamic signaling process. Prohibitions on formal transfers can either eliminate the signaling altogether (by preventing transfers from occurring) or they decrease the effectiveness of market signals by driving the transactions underground. This reduces the potential for gains, by allocating gains to those who have the ability to block the transfer (Segerfeldt 24).

New information regularly appears in the field of water use. Today, we know far more about the value of in-stream uses than in the past (Anderson and Snyder 111) and we can measure water quality both more accurately and in more dimensions. We also possess more information about the effect of water uses on downstream users; we can make more sophisticated measurements of the return flows from irrigation, leading to more accurate calculations about the impact of irrigation on downstream water users. There is potential for significant gains from trade when dynamic information about water is available to water users, and when that information can be incorporated into water use rights and contracts.

Market incentives to create new knowledge

Markets create opportunities for entrepreneurs to profit by developing new knowledge about both market conditions and goods and services. A market participant has the opportunity to profit by making use of newly-discovered information. As a result, markets encourage investment in information. As Segerfeldt notes, "A private firm competing with other firms for the customers' favor must always be devising new and better methods and must be as efficient as possible" (Segerfeldt, 24; Dietz & Vollebergh 1999, 339)

Applied to water, the incentive to create knowledge is likely to produce investment in four areas, all of which will serve to enable better use, production, delivery and conservation.

First, if holders of existing water rights are to trade those rights, then it is likely that entrepreneurs will invest in new technology to maximize the value of the rights. Indeed, where waste exists, so does an entrepreneurial opportunity (Morriss et al. 2002, 343). More efficient application of water in agriculture, for example, can free a portion of an existing allocation for new uses. Someone who invents a more efficient method of irrigation can thus create value for existing agricultural water users by allowing them to sell a portion of their water without reducing their agricultural production (Anderson and Snyder 11).

Second, entrepreneurs will invest in new sources of water, far beyond simply digging new wells. New sources of water exist, for only a very small percentage of the world's supply of fresh water is consumed by humans (Segerfeldt 14). Indeed, outside of a relatively few geographical areas, humanity faces only an economic scarcity – not a physical scarcity – of water (Segerfeldt 15; Anderson and Snyder 1–7). Potential water sources include reducing waste (Zilberman & Lipper 1999), recycling used water and developing greater understanding of hydrology. We could also allow water users who add water back into systems to receive credit for their inputs against their withdrawals.

Third, entrepreneurs will invest in alternatives to water use. Reducing use is not simply a matter of utilizing low-flow shower heads and front-loading washing machines. Water markets that fully price the use of water are likely to lead to less reliance on plants which require large amounts of water in landscaping in arid areas, for example. Similarly, other factors of production will be substituted for water in response to higher prices.

One stark example is the drastic reduction during the past century in the volume of water required to produce steel: in 1930, one metric ton of steel required 200 metric tons of water, today it requires an average of 20 metric tons. Indeed, some steel manufacturers today use only three to four metric tons of water (Segerfeldt 16; Anderson and Snyder Table 1.1, 9).

Similarly, agricultural uses vary considerably with respect to their demand for water. Farmers can readily shift to less water-intensive crops and more efficient irrigation techniques if they have the opportunity to profit from water sales (Anderson and Snyder 10–11; Zilberman and Lipper 1999, 147).[4] Water companies will reduce waste in transmission if water becomes more valuable (Segerfeldt 31–32). Producers and consumers "respond rationally to water prices" (Anderson and Snyder 11).

Finally, one of today's urgent challenges with respect to water is simply the lack of adequate investment in producing and distributing safe water (Segerfeldt 15). It is not that we lack knowledge of how physically to deliver water to those who lack it; we lack the knowledge of how to do so cost-effectively. Entrepreneurs who discover ways to reduce the cost of creating and maintaining the infrastructure necessary to produce and deliver clean, safe water can profit from so doing.

Markets enable diverse human ends

A significant advantage of markets over alternative means of valuing and allocating goods is that markets do not require a social agreement on the values of the uses to which people plan to put the goods they purchase in the marketplace. Compared to non-market alternatives, markets generate fewer conflicts over whether the appropriate ends have been selected as high value uses.

For example, agricultural and residential users often compete for the same water resources. Without a market to enable transfers, agricultural water rights holders whose rights are reallocated to residential uses are not compensated for their loss, while the residential users receive something of a windfall. Both groups are thus likely to expend resources in the attempt to, respectively, block or facilitate the reassignment (Anderson and Snyder 25–28).

Economists refer to this phenomenon as "rent-seeking". Rent-seeking leads to dissipation of welfare through political competition. The money spent to capture the gains or prevent the losses from the transfer – both of which are referred to as "rents" by economists – benefits no one (except the lobbyists and lawyers involved) and creates no value (Krutilla 1999, 252–253). Whether the transfer will occur depends in part on some socially agreed upon relative valuation of agricultural and residential uses. If the society is democratic, it is costly to use the political system to make allocations. It also increases conflicts between agricultural and urban interest groups. If the society is not democratic, the decision is possibly cheaper (it simply may be "what does the dictator think is the appropriate outcome?"). Decisions in this situation are even less likely to rely on appropriate criteria, for the decision will affect the regime's ability to maintain itself in power.

Summary

Markets for water rights hold the potential to unleash significant entrepreneurial activity in water production and consumption. We need many additional entrepreneurs, whether small or large, to focus on water at all levels because more than one billion people lack access to clean and safe water – a number which "has held constant for decades" (Segerfeldt 1). In this environment, water rights have the potential to evolve into a more complex bundle than has so far been recognized with limited efforts to create water markets (Anderson and Snyder 44).

Marketable water rights will acquire quality, temporal, and other characteristics, all of which would allow a wide range of transactions not currently possible. We cannot fully anticipate what these transactions will look like, because we lack all of the information necessary to do so. We do know that decentralized free markets can more effectively utilize information than even the most sophisticated administrative arrangement. Unless there are significant problems with markets, these advantages should create a presumption in favor of relying on markets.

CRITICISMS OF WATER MARKETS IN PERSPECTIVE

Markets provide consumers with a wide range of goods and services, including vital ones such as food and housing. As Anderson and Snyder note, "Water certainly is necessary for life. But clothing and shelter are also necessities, and there is no justification for having governments control their allocation" (49). The mere identification of a good as a necessity is thus an insufficient justification for removing it from the marketplace.[5]

The standard justification for use of non-market arrangements is the idea that 'market failures' will occur more readily, and at a greater scale, than failure by an intervening institution.

Are major market failures likely to occur with water markets? There are certainly many self-styled opponents of water markets (Shiva 2002; Barlow and Clarke 2002), including many economists (outlined in Bate and Tren 2002, 35). These critics' primary complaints about water markets concern the alleged existence of externalities, the existence of a natural monopoly in water provision, and issues relating to affordability (and justice) of requiring the poor to pay for water. Do any of these concerns suggest that an alternative to water markets is superior?

Externalities

Externalities exist when there are attributes of a good which are not reflected in its market price, and thus are not considered by decision-makers using market information (Wills 1997, 63). For example, suppose in-stream flows produce habitat for fish. Since the affected fish are unlikely to be market participants directly, the fish-habitat aspect of the in-stream flow characteristic of water may not be a characteristic for which there is market demand (Anderson and Snyder 19). Many opponents of water markets contend that such externalities are pervasive (ibid, 19–20; Segerfeldt 76; Shiva 2002, 6), while market proponents find that externalities "are overused as an argument for governmental intervention in water markets" (Anderson and Snyder 52).

However, the issue is not whether externalities exist, but the relative size of externalities which might result from alternative arrangements. Consider, for example, the destruction of the Aral Sea – perhaps the largest environmental disaster relating to water in modern times. This tragedy resulted from the misallocation of water resources by the Soviet bureaucracy – the antithesis of a market allocation (Bissell 2002, 41).[6]

There are several responses that get to the heart of whether claims about externalities in water markets have any merit.

First, there is a real question about the magnitude of such problems. In many instances, a good's un-priced attributes relate to other attributes which are priced. For example, game fish species are highly valued by sports fishermen and businesses which cater to those fishermen. Maintaining in-stream flows for sport fish also maintains the flows for non-sport fish. To the extent that sport- and non-sport fish are part of the same eco-system, the benefits of maintaining a healthy eco-system for the sport fish will provide 'spillover' benefits to other fish. The mere existence of a fish which is currently unvalued for human use is thus not sufficient to demonstrate that a market-based water allocation scheme would produce an externality.

Second, even if such a fish species existed, with attributes such that there was no market demand for providing it with sufficient water to survive, we need not do away with water markets to solve the problem. By introducing a market actor with a budget and a mission to protect such species, we can create a demand for this previously un-priced attribute without disrupting the other beneficial aspects of water markets.

Third, many external effects result from efficiency-enhancing transactions. So, for example, if Smith sells Jones some water rights, a third party (Black) may be harmed in the marketplace by Jones' new water rights. "Naturally the third parties would like to restrict competition" by prohibiting the transfer (Anderson and Snyder 85). However, such harms cannot be compensated – they are inherent to the competitive market process, and are thus a vital part of improving humanity's living standards.

Fourth, external effects may not be widespread. Agricultural communities often voice concern about large scale water transfers to urban users, usually based on the high prices that cities are willing to pay to acquire water rights. These concerns fail to consider the way that the units of water are likely to be valued: the first units of water are the most highly valued and likely to fetch a higher price, but it is unlikely that relatively high prices will be offered for all the units. Moreover, the low return many farmers receive for their water use is based on their current total use. Since farmers would sell the units which are of lowest marginal value, their returns would increase as the urban marginal demand falls, making it unlikely that as much water would shift as the initial price gap might suggest (ibid. 85–86).

These qualifications reduce the number of concerns about externalities, but they do not eliminate all concerns. Let us consider how significant these are likely to be.

Externalities are the (mostly unforeseen) consequences of economic activities, including many that are not subject to transactions at all – but could be. Economist Harold Demsetz explained that

Property rights specify how persons may be benefited and harmed and, therefore, who must pay whom to modify the actions taken by the persons. The recognition of this leads easily to the close relationship between property rights and externalities... What converts a harmful or beneficial effect into an externality is that the cost of bringing the effect to bear on the decisions of one or more interacting persons is too high to make it worthwhile (Demsetz 1969, 347–348).

How often is this true of water markets? It is certainly the case that there are aspects of water use which are not considered by the participants in a market transaction. For example, there may be unknown or unforeseeable impacts on an ecosystem when water is withdrawn at a particular time and place. By definition, therefore, no one in the market or in alternative institutions will be aware that such problems exist.

There may by other impacts, however, known to the scientific community but not to the market participants. Depending on the relative cost of doing so, entrepreneurs might respond to this situation by creating a new transaction – a new way of packaging water rights, for instance – which accounts for that information. Even when this involves disparate parties, there may be alternative contracting parties that solve the free rider problem (Anderson and Snyder 113–114; Bate 2003).[7]

All well and good, but sometimes such transactions may not occur because the transactions costs are simply too high. In these cases, markets will fail to provide a response capable of ensuring that all bad outcomes do not occur.

This is certainly true, but all institutions suffer from such problems. We cannot demand perfection from any institution or we are doomed to disappointment. The issue is whether markets fare better or worse when compared to the non-market allocation mechanisms touted as the solution to externalities (Bate and Tren 2002, 36).

Such solutions would occur "without the benefit of information contained in market prices" (Anderson and Snyder 21). As discussed earlier, however, markets excel at low cost information processing, relative to other institutions (Wills 1997, 42–45). Moreover, for a non-market institution to be preferable to markets in resolving an externality problem, the alternative must be less costly at solving the problem than the alternative of introducing a new market participant into the market (e.g. funding an organization of sporting enthusiasts to acquire in-stream rights).

The class of externalities where non-markets outperform markets would have to be either extremely high value, or quite numerous, to give a non-market alternative the advantage. However, the proponents of non-market arrangements have not provided a sufficient theoretical basis to set aside the presumption in favor of voluntary transactions in markets.

Natural monopolies

It is frequently claimed that water or various water projects constitute a "natural monopoly" (summary of literature in Anderson and Snyder 50–52). In this situation, supply of water would be controlled in each area by a single firm, which would then charge monopoly prices that harm consumers. (Somewhat paradoxically, market critics frequently propose a 'remedy' in the form of a state monopoly; for a critique of state monopolies see Robinson, this volume, chapter 8.)

This claim is flawed for two important reasons. First, the existence of a natural monopoly in any particular aspect of a water system does not justify the creation of a state monopoly on the entire system (Morriss 1998, 138–156). Suppose that the capital costs of water mains are prohibitively expensive, such that it is inefficient to provide multiple mains in a region. This fact does not justify creating a monopoly on water distribution. A wide range of potential strategies can be implemented to limit the natural monopoly to the specific geographical area where it exists. Delivery can be unbundled from supply (as was successfully done with natural gas) (Pierce Jr. 1994, 323–324) or contracts can provide incentives for appropriate performance by the operator of the natural monopoly aspect of the system (Morriss 1998, 184–186). At the same time, a local monopolist would (in order to maximize profits) have incentives to offer differential prices, ensuring that all consumers willing to pay the marginal cost of production are supplied. "Unlike the governmental and regulated alternatives, a private unregulated monopoly also would have strong incentives to hold down costs and supply an optimal quality of product" (Cowen and Brook Cowen 1998, 23).

Moreover, competition in water supply may be more widespread than many believe, because of the potential for competition between ground-water and surface water-based systems (Anderson and Snyder 1997, 51).

Second, the concept of a "natural monopoly" has come under critical academic scrutiny in recent years. More recent work has exposed serious conceptual flaws in the definition of natural monopolies, which casts doubt on the extent to which to which natural monopolies exist in the real world (Hazlett 1985, 6–7). The potential existence of a natural monopoly is thus an insufficient reason to reject the role of markets.

"Markets harm the poor"

Markets respond to resources and, by definition, poor people have fewer resources than those who are relatively better off. Many opponents allege that using markets to distribute water will make the poor worse off than an alternative (which does not rely on individuals' relative resource endowments). To make an accurate evaluation, we must consider the nature of the alternative institution: If water is not distributed via markets, how will it be distributed?

In the real world, administrative allocation systems tend not to favor the poor for the simple reason that the poor rarely possess enough political power to secure administrative allocations in their favor. Many state-subsidized water systems actually operate as a regressive tax, bestowing benefits on the wealthy while taxing the poor. So the appropriate benchmark necessary to make an accurate comparison is not a hypothetical "just" distribution which uses whatever ethical standard an opponent might favor, but the distribution which occurs with politically-determined, administrative resource allocations.

Nonetheless, with a market allocation of any good, the poor will get less than those who are relatively wealthier, simply because they have relatively fewer resources. The important question is whether they will obtain less water than in alternative, non-market arrangements?

The poor have actually fared comparatively well in market systems (Rosegrant and Gazmuri 1994); see also Southgate and Figueroa, this volume, Chapter 3). Real water costs for poor consumers have often dropped considerably when water systems compete to extend service to poor areas (Hinrichsen et al. 1998).

Moreover, privatization means that water systems significantly improve the quality of water which is delivered to poor neighborhoods. Where problems exist, in general the poor have been victims of incomplete markets rather than of overly aggressive extension of market forces.

By contrast, non-market water allocation mechanisms have not performed well generally or with respect to the poor. As Segerfeldt notes, "There are roughly as many extremely poor people in the world (people living on less than a dollar a day) as there are people without access to safe water. In fact, these are to a great extent the same people" (8–9).

Indeed, it is misleading to discuss "the poor" as if they were a monolithic group with identical interests. Many poor people in developing countries currently pay high prices for water because the only available clean water supply in their neighborhoods comes from water tankers or from distant wells (where the main cost is the time to fetch the water, and the inconvenience of not having piped water; both costs are generally not accounted for).

If markets result in water being delivered to poor neighborhoods by water mains, the cost of water– even if it is at a relatively higher price than the price for piped water in other neighborhoods – is still likely to be far less than the price they are currently paying. Even if opponents of privatization are correct when they claim that water markets increase the current price of piped water, the extension of water markets to poor areas is still likely to benefit the poor – especially through improved health and sanitation, convenience and competition.

Claims that markets cause the poor to suffer are thus underspecified, because the impacts of markets are likely to vary among groups within the larger population of "the poor." For many poor people around the world, water markets are likely to produce a net gain. At the same time, if an increase in price did negatively impact the poor, this could be offset by some kind of voucher or direct transfer, such as Chile's use of water stamps.

The alternatives to markets

For those who claim to identify a problem with markets, we must ask "Compared to what?" We cannot judge institutions in isolation; we must compare the strengths and weaknesses of alternative institutions before we can draw conclusions about any particular arrangement.

Fortunately we need not consider a wide range of alternatives to understand the institutional strengths of markets. All of the alternatives to markets share one of two common characteristics: they rely either upon administrative allocation of water rights or they are characterized by an absence of property rights in water. These two characteristics enable a comparison to be made between markets and a wide range of alternatives.

Rights-based vs. non-rights based allocations

In non-rights-based allocations, water belongs to the person who takes it. Without rights to water, it must remain in the commons until captured and consumed. (Moreover, if even capture does not convey rights, the water must be consumed rapidly, or it will be subject to capture by another.) Garrett Hardin explained non-rights-based allocation systems in his classic article *The Tragedy of the Commons*, which argues that resources left in the commons would inevitably and disastrously be over-exploited (Hardin 1968). The solutions Hardin considered were either privatization (e.g. the creation of rights) or "coercive laws or taxing devices" to prevent over-exploitation of the commons (ibid.).

Fortunately, subsequent analyses found that the tragic circumstances postulated by Hardin rarely exist in the world, precisely because their consequences are so disastrous (Ostrom 1990). Resources that may appear to be a "commons" are often regulated by systems of customary rights which operate as the functional equivalent to property rights. Thus in Hardin's paradigmatic case of the English common pasture, later research revealed the existence of customary institutions which limited the ability of individuals to add cattle to the common field. Here, and with many other resources, customary institutions prevented the "tragedy of the commons" from occurring.

In instances where a rule of capture applies, however, the tragedy of the commons does occur. For example, during the 1990s, the groundwater regime in Texas tended toward a tragedy of the commons because it relied on the rule of capture and did not allow rights in ground water in place (Thompson Jr. 2000, 266–267; Zilberman & Lipper 1999, 151).

Opponents of water markets often assert that "community based" institutions are superior to markets (Shiva 2002, 19). Yet they lack explanations as to how such institutions create incentives for appropriate water use, and they are selective in their choice of communities.[8]

It may be legitimate to be concerned about who receives the initial allocation of rights, but once rights are allocated, nothing prevents the rights-holders from creating community-based common pools out of their rights. Indeed, some communities have managed just such arrangements.[9] Given the specificity of the cultural arrangements necessary to make such systems function, however, there seems little reason to force everyone to adopt such systems.

The lesson of the tragedy of the commons is that rights-based solutions have an important advantage over non-rights based institutions when we wish to protect a particular resource from overexploitation. Since water falls into this category of resources, a rights-based solution will (absent extraordinary circumstances) create better outcomes than a non-rights based solution.

Administrative vs. voluntary allocations

If we choose a rights-based allocation as the basis for our system, the remaining issues are how to allocate the initial distribution of rights, whether to allow reallocations of rights once they are allocated, and, if reallocations are allowed, whether voluntary reallocations may occur between individuals or whether these require approval by a centralized decision-maker. To some extent, errors in the initial distribution can be corrected through transfers, if they are allowed.

Thus the question revolves around whether or not to allow voluntary transfers of rights, or whether to limit voluntary exchanges of rights through some administrative mechanism.[10] Voluntary transfers imply that the transaction will produce a net gain in wealth for the parties; by definition, people do not engage in voluntary transactions that they believe will make them worse off.

The same cannot be said of involuntary transfers, however, since the only evidence is that the token administrative authority decided that the transfer was desirable.[11] Since the authority may possess motives which are less than desirable – a possibility in all systems short of a government staffed by angels (Federalist No. 51) – the mere existence of the transfer does not tell us that it increases net value.

If we allow voluntary transfers in a rights-based system, we have a market. If we permit only administrative reallocations to occur, no market exists. There are various intermediate regimes (in which rights might be taken for public use, for example, with compensation paid), but a comparison of these two extremes is useful.

Non-market allocations require administrators to collect a great deal of essential information. For example, an administrative decision about how water should be allocated among competing agricultural users must account for the relative costs and quality of the land, and its suitability for irrigation, as well as the users' plans for which crops to grow. A more complex issue would be a trade-off between urban and agricultural uses, a decision which would require an even wider range of facts to make a "correct" allocation (under whatever standard is used). The answers to such questions are neither obvious nor inexpensive to discover.

The key to this discussion is that administrative allocations require considerable information. It was precisely this point that led Nobel laureate economist Friedrich Hayek to make a convincing case for the superiority of markets over central planning generally. Hayek explained that markets are an efficient means to process decentralized information, but central planners are not (Hayek 1945; Wills 1997, 42–43). The greater the cost of incorporating the information we desire from the planner, therefore, the less likely it is that the planner will use it to inform allocation decisions.

Advocates of non-market mechanisms make two assumptions. First, they assume that such mechanisms are superior to the market because they incorporate some information or value into the decisions that markets cannot capture, or have not captured. Second, they assume that this superiority is sufficiently large that it offsets the market's superiority in considering local knowledge. These are dubious assumptions, since the transaction costs involved in amassing and processing information (e.g. the plans of individuals, local knowledge of specific characteristics of individual plots of land, etc.) are significant.[12]

Moreover, they make the unrealistic assumption that the administrative mechanism will be designed so as not to incorporate inappropriate considerations into the decision making (e.g. political factors) (Anderson and Snyder 59–60). The transactions costs of "accountable government" are themselves considerable. Due process and separation of powers, for example, impose considerable transactions costs (Wills 1997, 108–109) and many countries lack even these basic elements of the rule of law. In combination, these unrealistic assumptions suggest that non-market alternatives are unviable.

Foundations for water markets

Markets generally require secure property rights and the rule of law to function (Anderson and Snyder 1997, 14). Water markets are no different.

What does it mean to provide secure property rights to water? Property rights must be "3-D" to support markets: definable, defensible, and defeasible (Yandle and Morriss 2001; Wills 1997, 23). For property rights to water to be *definable* requires that we define the characteristics of water which make up the bundle of rights. Early definitions of water rights in the American west focused on quantity and priority; today we also recognize the importance of a wide range of quality variables. *Defensible* property rights are rights which are recognized by law and capable of being defended in court against unauthorized interference by others. *Defeasible* property rights are rights which can be transferred to others.

The definition of rights for a commodity such as water presents challenges for the legal system, since the relevant dimensions are likely to change over time as our knowledge of water and our thirst for transactions involving water rights increase. The contours of water rights must therefore remain flexible, allowing courts to expand their boundaries as new knowledge is brought to bear. Thus, a common law process for defining water rights will likely be superior to a statutory definition, because the flexibility of the common law will make the rights more readily expanded to incorporate new dimensions.

A clear definition of water rights is most important: "Well-defined water rights give individuals a clear idea of what they own" and hence they define the realm of actions that are possible with this resource (Anderson and Snyder 23).

Unfortunately, water rights in the United States have often fallen victim to outdated knowledge through statutory definitions which fail to allow for expanding definitions of the dimensions of the property right. In-stream uses, for example, were forbidden by many western states which utilized the prior appropriation doctrine, which followed a "use it or lose it" approach and did not recognize preservation of stream flow as a "use" (ibid., 80–81). Similarly, transactions are obstructed by definitions of water rights that tie ownership of the rights to ownership of a particular parcel of land. This increases the transactions costs of a transfer, since the buyer of the water must also purchase the land associated with the water.

WATER MARKETS AND THE FUTURE

Water can be precious or worthless – it depends on the time and place where it sits and flows. In arid regions, water (when present) tends to have higher values than in more humid areas, where flooding can give water a negative value in many instances. The value of water thus depends on a wide range of factors, a range that is difficult for human institutions competently to value outside the context of real choices by real people with real resources. All too often, administrative allocations of water freeze the definition of water, neglecting the fact that our knowledge of the value of water is dynamic.

Today, we know far more about water institutions, the attributes of water and its possible uses than we knew fifty or one hundred years ago. We also have ideas about how water might be used in the future. Yet we cannot say with any certainty what the most valued use of water will be tomorrow. In 1900, who would have predicted explosive growth in Phoenix, Arizona (in the middle of the desert) since World War II, the development of shrimp farming in arid regions in Texas, or the expansion of an extensive sport fishing industry throughout the Rocky Mountain west? Given all that we do not – and cannot – know about the future, there is a high premium on the adaptability of institutions that humanity chooses to use for the management of our most precious resources.

Fortunately, there is a simple, straightforward and inexpensive solution to the economic scarcity of water that plagues more than a billion of our fellow humans: we must clear away "the current morass of legislative and administrative rules" that obstruct the development of water markets (Anderson and Snyder 13). By enabling such markets to develop, we will unleash the enormous creative power of entrepreneurs to address the problems of inadequate water, inefficient uses of water, and poor water quality.

The obstacle to unleashing the value of water is that special interests do attempt to manipulate the state to protect their claim to this increasingly valuable resource. In an earlier history of the evolution of water law, I noted that "It was not the manipulation of water but the manipulation of government about water that marked the West" (Morriss 2001, 867). Attempts to rely on "cultural" explanations for water use neglect the fact that it is not an inherent characteristic of a culture that determines water use, but the incentives created by that culture's institutions (Shiva 2002, 119).

What we do know is that policies matter in human terms. As Fredrik Segerfeldt noted, "there is actually an astonishing level of difference between countries at similar levels of development, suggesting that policies matter a lot ... " (Segerfeldt 15). We have spent decades and billions of dollars on attempts to provide water without markets. Is it not time to try something else?

Notes

1. A more extensive version of this chapter is forthcoming in the proceedings of the Texas Tech University Center for Water Law & Policy's first annual conference on water law, *The Value and Ethic of Water*, under the title "Real People, Real Resources, & Real Choices: The Case for Market Valuation of Water," *Texas Tech Law Review*, vol. 38 (forthcoming 2006).
2. One caveat to the above is needed: I am assuming that the transactions which occur as a result of the reduced information costs made possible by markets do not themselves produce *compensable* harms for others. See Anderson and Snyder (1997, 140–141), discussing need to limit lawsuits to instances of actual damages. This is assumption is reasonable because if the transactions produced compensable harms, the victims would seek compensation. The restriction of concern to compensable harms is important, for there are a variety of harms which transactions may cause for which there is not only no compensation due but whose occurrence is not recognized as a loss. These include losses in the marketplace as a result of competition by others. If a competitor opens a store down the street from me and offers lower prices or better service and so lures away my customers, I have been harmed but the law does not recognize that harm as compensable. (In a section below, I discuss the externality-based critique in more detail.)
3. As Richard Stroup summarizes:
   Entrepreneurship is the key to efficiency in a world where technology and relative prices change rapidly. Private firms are constantly adjusting their own organizations to handle changing problems and oppor-

tunities. For them the carrot is profit; the stick is failure to survive under competition. Feedback to them from the product market tends to be constant (Stroup 2000, 485, 489).

4. Shiva argues that "[b]y failing to recognize water as a limiting factor in food production, industrial agriculture has promoted waste" (Shiva 2002, 108). This claim, like many made in the anti-market literature, rests on a profound misunderstanding of current water institutions. What Shiva is observing is more likely the absence of markets, in which politically powerful agricultural interests have secured subsidized water. For example, Shiva cites the example of "sugar barons" in Maharashtra opposing the diversion of water from sugar production (ibid. 125). Of course the sugar barons are not operating in a market when they threaten that "a canal of blood will flow" if they do not receive their water, for there is no market mechanism to create "canals of blood."

5. Moreover, those who wish to remove one necessity from the marketplace but not others bear the burden of identifying a principled distinction that explains the different results.

6. Bizarrely, Shiva cites the Aral Sea disaster as a criticism of markets (Shiva 2002, 111–112). Her argument appears to be that the problem arose because of Soviet "industrial farming" and that such farming is somehow a market phenomenon (ibid. 112). The Soviet economy, however, was not a market economy. Of course, a market could never produce the Aral Sea disaster because the cost of acquiring the rights necessary through voluntary transactions would far exceed the benefits of the project. Only a government can create a disaster of such a magnitude for only a government can seize property rights on such a scale without paying compensation.

7. Ducks Unlimited is an example of a successful private effort to save sufficient wetlands to use for migration purposes ensure migratory waterfowl can successfully migrate. Similar groups include Trout Unlimited and The Nature Conservancy. See Anderson and Snyder (1997, 113–114). Similarly English fishermen banded together in associations to protect water quality (Bate 2003).

8. Shiva, for example, is highly critical of "cowboy economics" and the prior appropriation doctrine in the American West (yet she is seemingly ignorant of its roots in earlier legal systems). She fails to recognize that the development of prior appropriation was a

community response to arid conditions in the West (Morriss 2001, 884–88). At the same time, she praises communal water systems in India, which were based on the existence of a landless caste whose neutrality in allocating water stemmed from their lack of resources. Shiva describes the system as follows:

> To ensure neutrality, nirkattis were chosen from the landless caste – the Harijans – who were granted autonomy from landowners and caste groups. Only the Harijans held the power to close and open the tanks or vents. Once the farmers laid down the rules of distribution, no individual farmer could interfere and those who did could be fined. This protection of the associations from the economically powerful ensured water democracy (Shiva 2002, 30).

Shiva does not explain how the existence of a "landless caste" is consistent with democracy nor how the Harijans would maintain their neutrality if they were allowed to own land. It is difficult to see the moral merit in a system which depends upon mass disenfranchisement.

9. Symposium discussion noted several water systems in New Mexico pueblos were excellent examples. Symposium, "The Value and Ethic of Water", Center for Water Law and Policy, Texas Tech University, Lubbock, Texas, November 2005.

10. I assume that some transfers will be permitted by the system, whether voluntary or involuntary, because there are very large dead weight losses from forbidding transfers entirely (Anderson and Snyder 1997, 60).

11. For example, as Anderson and Snyder (1997) note,

> Rents can also be obtained ... by redistributing the economic pie using the coercive powers of government. Suppose, for example, that the government is willing to tax a segment of the population to provide a subsidy for an irrigation project that will allow grape production. If water is free to the grape farmer but costs taxpayers $60 per acre-foot and produces additional grapes worth $50, the grape farmer is clearly better off. Those who obtain the subsidized water from the project capture rents equal to the difference between what they have to pay and what the water is worth to them. In that process, taxpayers lose part of their income, but grape growers receiving the subsidized irrigation water enjoy an increase in their wealth (48).

12. The existence of uncertainty in the neoclassical paradigm is often used as a justification for rejecting markets. As Anderson and Snyder (1997) note:

    > *some economists are skeptical about markets because individuals must act in a world of uncertainty without perfect information. In the traditional economics paradigm, uncertainty and information costs imply market failure that warrants government intervention. For example, in the presence of uncertainty about water availability, many states have drought plans ready to be implemented by "water experts" in state agencies. Economist Thomas Sowell notes that "the conduct of social activities depends upon the special knowledge of the few being used to guide the actions of the many"* (18–19).

    Such an analysis neglects the important incentive effects of markets and the role of entrepreneurs, as stressed by Anderson and Snyder. See also Anderson and Snyder (1997, 24); Segerfeldt (2005, 71). Some water market critics fail to grasp the local nature of knowledge that markets allow to be considered. Shiva, for example, argues that "cooperative management systems" are superior to markets because they are free from control by "dominant bureaucracies" (Shiva 2002, 123). Such a claim may have been true historically, and if so the cooperative management system is possibly the equal of a local market in its ability to take into account local information. There are multiple examples of institutions that manage resources in specific locations through customary law and other means; see Ostrom (1990). See also Wills (1997, 99–100). Such institutions are not inconsistent with markets, and indeed rely on markets for information about events outside the local community.

References

Anderson, Terry L. and Snyder, Pamela. (1997). *Water Markets: Priming the Invisible Pump*, Washington, DC: Cato Institute.

Barlow, Maude and Clarke, Tony. (2003). *Blue Gold*. London: Earthscan.

Bate, Roger. (2003). 'Saving Our Streams: The Role of the Anglers' Conservation Association in Protecting English and Welsh Rivers', *Fordham Environmental Law Review*, vol.14, no.2 (Spring), p. 375.

Bate, Roger and Tren, Richard. (2002). *The Cost of Free Water: The Global Problem of Water Misallocation and the Case of South Africa.* Johannesburg: Free Market Foundation.

Bissel, Tom. (2002). 'Eternal Winter: Lessons of the Aral Sea Disaster', *Harper's*, April, pp. 41–56.

Brook Cowen, P. and Cowen, T. (1998). 'Deregulated private water supply:. A policy option for developing countries'. *Cato Journal*, vol.18, no.1, 21.

Demsetz, Harold. (1967). 'Toward a Theory of Property Rights.' *American Economic Review* vol. 52 (May), 347.

Dietz, F.J. and Vollebergh H.R.J. (1999). Explaining Instrument Choice in Environmental Policies. In J. van der Bergh (Ed.) *Handbook of Environmental and Resources Economics* (p. 339). Cheltenham UK: Edward Elgar.

Federalist No. 51. (1788). *The Federalist Papers* by Alexander Hamilton, James Madison and John Jay.

Grimmelmann, James. (2005). 'Regulation by Software', *Yale Law Journal*, vol. 114, 1719.

Hahn, Robert W. and Hird, John A. (1991). 'The Costs and Benefits of Regulation: Review and Synthesis', *Yale Journal on Regulation*, vol.8, 233.

Hardin, Garrett. (1968). 'The Tragedy of the Commons', *Science*, vol.162, 1243.

Hayek, Friedrich A. (1945). 'The Use of Knowledge in Society', *American Economic Review*, vol. 35, p.519.

Hazlett, Thomas. (1985). The Curious Evolution of Natural Monopoly Theory. In Poole, Robert W. Jr (Ed.) *Unnatural Monopolies: The Case for Deregulating Public Utilities.* Lanham, MD: Lexington Books.

Hinrichsen, D., Robey, B., and Upadhyay, U.D. (1998).'Solutions for a water-short world.' *Population Reports* 26.

Kreitner, Roy. (2000). 'Speculations of Contract, or How Contract Law Stopped Worrying and Learned to Love Risk', *Columbia Law Review*, vol.100, 1096.

Krutilla, Kerry. (1999). Environmental Policy and Transaction Costs. In J. van der Bergh (Ed.) *Handbook of Environmental and Resources Economics* (p. 249). Cheltenham UK: Edward Elgar.

Macey, Jonathan R. (1994). 'Administrative Agency Obsolescence And Interest Group Formation: A Case Study of the Sec at Sixty', *Cardozo Law Review*, vol.15, 909.

Morriss, Andrew P. (1998). 'Implications of Second-Best Theory for Administrative and Regulatory Law: A Case Study of Public Utility Regulation', *Chicago- Kent Law Review*, vol. 73, 135.

Morriss, Andrew P. (2001). 'Lessons from the Development of Western Water Law for Emerging Water Markets: Common Law vs. Central Planning', *Oregon Law Review*, vol. 80, 861.

Morriss, Andrew P.,Yandle, Bruce, and Anderson, Terry L. (2002) 'Principles for Water', *Tulane Environmental Law Journal* vol. 15, 335.

Ostrom, Elinor. (1990). *Governing the Commons: The Evolution of Institutions for Collective Action.* Cambridge UK: Cambridge University Press.

Ott, Harry. (2005). Comments at Symposium, "The Value and Ethic of Water", Center for Water Law and Policy, Texas Tech University, Lubbock, Texas, November.

Pierce Jr., Richard J. (1994). 'The State of the Transition to Competitive Markets in Natural Gas and Electricity', *Energy Law Journal*, vol.15, 323.

Rapier, John. (2005). Comments at Symposium, "The Value and Ethic of Water", Center for Water Law and Policy, Texas Tech University, Lubbock, Texas, November.

Rosegrant, M.W. and Gazmuri-Schleyer, R. (1994). *Reforming Water Allocation Policy Through Markets in Tradable Water Rights: Lessons from Chile, Mexico, and California.* Washington DC: International Food and Production Technology Division, EPTD Discussion Paper No. 6.

Segerfeldt, Fredrik. (2005). *Water for Sale: How Business and the Market can Resolve the World's Water Crisis*. Washington, DC: The Cato Institute.

Shavell, Steven. (1987). 'Economic Analysis of Accident Law', *NBER Working Papers*. Working Paper 9483 (May), p. 186.

Shiva, Vandana. (2002). *Water Wars: Privatizations, Pollution and Profit.* Cambridge, Massachusetts: South End Press.

Stroup, Richard L. (2000). 'Free Riders and Collective Action Revisited,' *The Independent Review*, vol. 4, no.4, p. 485.

Surowiecki, James (2004). *The Wisdom of Crowds: Why the Many Are Smarter Than the Few and How Collective Wisdom Shapes Business, Economies, Societies and Nations,* New York: Doubleday.

Thompson Jr., Barton H. (2000). 'Tragically Difficult: The Obstacles to Governing the Commons', *Environmental Law* vol. 30, p. 266.

Thomsen, Esteban F. (1994). Prices and Knowledge. In Boettke, Peter J (Ed.) *The Elgar Companion to Austrian Economics* (p. 169). Cheltenham, UK: Edward Elgar Publishing.

Wall Street Journal (2005). 'Southwest Air's Profit Soars as Jet Blue's Descends'. Susan Warren and Susan Cary, Oct. 21.

Wills, Ian (1997). *Economics and the Environment: A Signalling and Incentives Approach*, Sydney: Allen & Unwin.

Woods Jr., Thomas E. (2005). *The Church and the Market: A Catholic Defense of the Free Economy*, Lanham, MD: Lexington Books.

Yandle, Bruce and Morriss, Andrew P. (2001). 'The Technologies of Property Rights: Choice Among Alternative Solutions to Tragedies of the Commons', *Ecology Law Quarterly*, vol. 28, p. 123.

Zilberman, David and Lipper, Leslie. (1999). The Economics of Water Use. In J. van der Bergh (Ed.) *Handbook of Environmental and Resources Economics* (p. 141). Cheltenham UK: Edward Elgar.

3 Reforming water policies in Latin America: Some lessons from Chile and Ecuador

Douglas Southgate and Eugenio Figueroa B.

As a sub-discipline, environmental economics is something of an outlier. Whereas economics as a whole focuses on the efficiencies realized if specialization and trade are unencumbered – market functionality, as one might say – economic analysis of natural resource issues is mainly about market failure, such as the inefficiencies that result if a competitive industry treats its adverse environmental impacts as externalities.

To be sure, externalities (both positive and negative) are ubiquitous. Consider the downstream consequences of natural resource use in the upper reaches of a drainage basin. If agricultural chemicals are misapplied or if hillsides are over-plowed, then water is contaminated and sediments accumulate at lower elevations. In contrast, establishing forested strips alongside streams creates important hydrological benefits for downstream populations. As is made clear in any textbook on environmental economics, of which dozens have been published, negative spillovers are excessive because costs are not fully internalized by responsible parties. By the same token, positive externalities are under-produced because suppliers do not capture all benefits.

While market failure is an important cause of environmental deterioration, one must keep in mind that market allocation of natural resources can be efficient, and indeed often is. Moreover, non-market allocation does not guarantee desirable outcomes. To the contrary, the latter approach frequently creates inefficiency, inequity, and damage to the environment.

Non-market allocation of water is an excellent case in point. In nearly every part of the world where water is provided by the public sector, the prices that households, farms, and other consumers pay for this essential resource fall short of the true cost of building, operating, and maintaining the dams, canals, and other infrastructure required for reliable supplies. High subsidies – or, if one prefers, poor cost-recovery – cause waste and misallocation on a grand scale. The beneficiaries of this inefficiency are mainly those who are wealthier – and politically better connected – while the losers are the poor. Furthermore, poor cost-recovery in potable-water, irrigation, and other systems reduces funding for the protection of water sources and the containment of pollution. Thus, the environment suffers.

Removing price distortions may not solve all of the world's water problems. However, supplies of the vital liquid will never be adequate for all humanity in the absence of pricing reform. Distortions in the prices of potable water have been reduced recently in Latin America, frequently (though not always) due to privatization. However, attaining efficient pricing in the irrigation sector, which accounts for a large share of total water use and consumption,[1] is a greater challenge. The problem is that real-estate values, which are highly sensitive to subsidies, tend to fall as cost-recovery improves in irrigation projects. Appreciating this linkage full well, farmers who benefit from under-priced water are seldom enthusiastic about pricing reform.

The disadvantages of subsidized prices and the benefits of reform: potable water in Quito

The inefficiencies, inequities, and negative environmental consequences of under-priced water were all on full display in the capital of Ecuador during the late 1980s. The municipal water company rarely recovered more than 50 percent of its costs. Chronically short of funds, the company did not serve its customers (i.e., those with pipes running into their homes and businesses) well; during some parts of the year, for example, water-supply interruptions occurred daily in many sectors of the city. Even worse, financial constraints prevented the firm from serving all its potential customers;[2] approximately 35 percent of the metropolitan population was unconnected to the municipal system, and instead relied on water delivered by tanker trucks (Carrión 1993). Since this mode of delivery is relatively expensive, prices paid by those households without a piped connection were nearly ten times the prices charged to households who were fortunate enough to be served by the municipal company (Southgate and Whittaker 1994, 72–73).

The neighborhoods which depended on tanker water were generally poorer than the rest of the metropolitan area. Many of these neighborhoods – like the slums which surround many Latin American cities – were formed when squatters had invaded farms on the outskirts of Quito. These new suburban communities petitioned for public utilities, but they lacked political clout and hence waited endlessly to be served. So in the late 1980s, Quito's potable-water subsidies created stark inequities, just as under-pricing has done in metropolitan areas throughout the region before, during, and since. The policy tended to deny access to piped water to the poorest segment of the urban population, which then relied on much more expensive sources (ibid., 73).

Poor cost-recovery has been inequitable in another respect. The subsidies required to fund below-cost supplies exacerbate fiscal deficits. In Ecuador and many other Latin American countries, governments have printed more money to pay for their debts. The resulting inflation, which essentially acts as a stealth tax on savings, is particularly burdensome for the poor, since their earnings tend not to adjust fully to higher living costs and because rising prices tend to ravage any savings they possess. At the end of the day, the poor derive little benefit from subsidies – including poor cost-recovery in potable-water systems – while bearing much of the burden in the form of an inflation tax.

Finally, Quito's policy of under-pricing water during the late 1980s was impossible to justify on environmental grounds. The municipal company struggled to pay for the construction and upkeep of its pipes, pumps, and other infrastructure. As such it had little funding for watershed conservation, which is crucial for averting seasonal and other water shortages. Likewise, nothing was done about wastewater management (beyond channeling untreated sewage into nearby rivers) due to a lack of financial wherewithal.

Potable-water subsidies being inefficient, inequitable, environmentally damaging, and ultimately unsustainable, cost-recovery increased substantially in Quito during the 1990s. By the middle of the decade, prices were generally in line with the cost of water-delivery. A better financial situation allowed the municipal company to provide more reliable service to existing customers (e.g., by reducing the frequency of interruptions), which explains why the public accepted an increase in prices. The company also undertook major expansions of the system which have been enormously beneficial for newly-served neighborhoods whose residents no longer rely extensively on water delivered by tanker-trucks. By 1993, the share of the metropolitan population provided direct service by the water company had risen to 80 percent (Carrión 1993) and was approaching 90 percent at the end of the 1990s.

In Argentina (see Box 3.1), Brazil (Fujiwara 2005), and other parts of Latin America, the extension of municipal systems into impoverished neighborhoods – as happens if privatization or some other reform improves the efficiency with which potable water is delivered – has had a positive impact on public health, although there is limited documentation of this impact in Quito. However, the Ecuadorian capital can point to demonstrable progress on the environmental front. A water fund (FONAG) was created to fund watershed conservation. In addition, wastewater treatment facilities are being constructed for the first time ever in the city. Regrettably, progress on both these initiatives is being stymied because cost-recovery, after improving significantly during the 1990s, has deteriorated in recent years.[3]

Box 3.1 **Public health benefits of potable-water privatization in Argentina**

During the 1990s, when the Argentine government sold off a number of state-owned enterprises, legal changes were implemented that allowed for privatization of municipal water companies. Approximately 30 percent of these companies were sold to private operators, while the rest remained in the hands of local government. This change in ownership created a real-world experiment which could potentially reveal the impacts of potable-water privatization on public health. Analyzing data from this experiment, Argentine economist Sebastián Galiani and colleagues Paul Gertler and Ernesto Schargrodsky found that these impacts are positive, especially for the poor.

Since children are especially vulnerable to infectious and parasitic diseases that are contracted as a result of consuming contaminated water, the three economists selected child mortality as the dependent variable in their analysis. In addition, major factors other than water privatization that might have affected mortality – such as per-capita spending on social services by the local government – were included in the statistical model.

The economists' findings comprise compelling evidence that public health improves as potable water systems are privatized. Figure 3.1, taken from their study in the *Journal of Political Economy*, indicates that the trend in child mortality was uniform throughout Argentina through 1995 (when privatization started to occur). At this point, there was no significant difference between areas served by water companies that were later privatized, and areas served throughout the decade by companies controlled by local authorities. The same exact trend continued, neither accelerating nor decelerating, during the second half of the decade in non-privatized settings. But meanwhile, the trend toward lower mortality rates accelerated in privatized settings, which caused statistically significant differences to emerge between the two areas.

Galiani, Gertler, and Schargrodsky estimate that, all else remaining the same, privatizing potable-water delivery results in an 8 percent decline in child mortality. Perhaps because several privatization agreements stipulated that new owners must extend service into poor neighborhoods, the benefits of privatization were found to be especially great there. Indeed, they estimate that privatization reduced child mortality by 26 percent in the poorest areas.

Figure 3.1 **Trends in child mortality in privatized and non-privatized water companies**

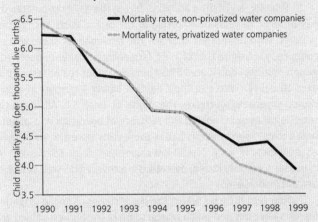

Note: There is no precise date of the shift from public to private with the companies involved, but the shift in ownership did occur during the mid-1990s. It is after 1995 (dotted line) when the trend toward lower childhood mortality in privatized settings was observed.
Source: Galiani, Gertler, and Schargrodsky (2005), p.86.

To ensure that these estimated impacts are not spurious, the three economists investigated different causes of mortality. Potable-water privatization was found to be significantly related to fewer deaths from infectious and parasitic diseases. In contrast, there is no correlation between privatization, on the one hand, and mortality having nothing to do with water conditions, on the other. At least in Argentina, then, the conclusion that privatizing the delivery of potable water benefits human health, especially the health of poor people, seems beyond dispute.

Source: Galiani, S., P. Gertler, and E. Schargrodsky (2005).

The disadvantages of irrigation subsidies and impediments to pricing reform: the case of Ecuador

The adverse impacts of poor cost-recovery in potable-water systems correspond directly to the consequences of irrigation subsidies, although the gap between the prices farmers pay and the expense of channeling water to their fields tends to far outweigh levels of subsidization in municipal systems.

In Ecuador, legislation adopted during the 1970s mandates the pay-off over 75 years of a portion of the capital costs of irrigation projects, with no interest charges. The remainder of these costs is completely subsidized. As of 1989, farmer-beneficiaries of public irrigation projects contributed $1.57 per hectare for the infrastructure used to deliver water to their fields – barely 1 percent of actual capital expenses (Southgate and Whitaker 1994, 62). Although the same legislation mandated full recovery of operation and maintenance costs, approximately half of these costs were also subsidized. As irrigation is very capital-intensive, and since 99 percent of expenses were subsidized, overall cost-recovery was well below 10 percent (*ibid*., 64–65).

Paying very little for water, the beneficiaries of public irrigation projects have not had much reason to make the highest valued and best use of hydrological resources. For example, on-farm water efficiencies[4] are very low in Ecuador. For this and other reasons, the return on governmental investment in irrigation has been disappointing, with benefits (which comprise the market value of additional production) equaling no more than half the costs of irrigation during the 1980s (*ibid*., 65–66).

Even though benefits fall well short of costs, subsidized irrigation is highly profitable for its farmer beneficiaries. A hypothetical example illustrates the point. For instance, assume a farmer pays 10 percent of total irrigation costs. Assume also that the value of additional crop production – which is captured entirely by the farmer – is equivalent to 30 percent of those same costs, after an allowance is made for expenditures on other agricultural inputs. In this example, society suffers a loss – the 70 percent of irrigation expenses that is not compensated for by any benefit. However, for the farmer, three pesos come his way for every one peso he spends on water.

This sort of profitability explains why farmers not only seek to maintain existing subsidies, but lobby for new irrigation projects as well. Responding to this political pressure, the Ecuadorian government constructed new systems during the 1980s which had a combined capital cost of US$1.1 billion. Likewise, other projects being designed or given serious study during that same decade had an aggregate capital cost of about US $1.1 billion (*ibid.*, 61–62). To put these expenditures into perspective, one needs to keep in mind that Ecuador's national debt in the late 1980s stood at $12.0 billion and its GDP was a little over US $10 billion.

In the early 1800s, classical economist David Ricardo showed that the main effect of subsidizing agriculture is to drive up the value of rural real estate (Ricardo 1965, 33–45). This observation certainly applies to poor cost-recovery in irrigation systems. In Ecuador, price-premiums for irrigated land have been found to range from a little less than US $400 to nearly US $4,000 per hectare and to average nearly US $1,100 per hectare (Southgate and Whitaker 1994, 65). In contrast, non-irrigated farmland was regularly exchanged for US $500 per hectare or less.

The price-premiums for irrigated real estate mainly benefit well-off farmers, and not the rural poor. Indeed, inequitable distribution of the gains from water resource development is all but inevitable. Individuals with good political connections (not least, politicians themselves) are the first to find out about new projects. Before a new irrigation initiative has been publicized, these favored individuals buy land in the area which will be affected, knowing that real-estate values will rise substantially once the initiative is announced. Precisely because of this sort of transaction – as well as the ability of these same people to obtain cheap, government-supplied water for land they have already owned for a long time – the benefits of irrigation are concentrated among the privileged elite, rather than being distributed widely among the rural population.

Finally, irrigation subsidies, like poor cost-recovery in potable-water systems, lead not only to inefficiency and aggravated income disparities in the countryside but also to environmental damage. With the budgets for water resource development used almost entirely to subsidize existing projects and to construct more of the same, technical assistance on the proper use of irrigation water and the management of irrigated soils is chronically modest. In addition, national and regional agencies responsible for water resource development, which have a clear mandate for watershed conservation and pollution-control, have little funding to spare for these activities. Given these conditions, natural resources are wasted, polluted, and otherwise degraded.

In light of all its manifest disadvantages, subsidized irrigation has proven to be remarkably durable – certainly more durable than the policy of under-priced potable water. Modest progress toward rational pricing made in Ecuador during the 1990s was made because of an initiative (underwritten by the World Bank and Inter-American Development Bank) to create local water users' associations. Each of these associations was given exclusive responsibility for operating and maintaining a local irrigation system that had been built by the public sector. Rather than receiving government monies, the new entities were to be financed entirely by fees collected from members.

This initiative, which is similar to the approach followed in a number of Latin American nations, works insofar as devolution of the responsibility for irrigation management improves the quality of operations and maintenance, which in turn enhances farmers' acceptance of higher water prices. So far in Ecuador, a handful of old public projects are now being managed by local associations.

Compared with diminished governmental subsidization of operation and maintenance, which the devolutionary approach has facilitated, a large share of capital costs is unlikely ever to be recovered. Inefficiency is part of the problem. As already indicated, the benefit-cost ratio for public irrigation in Ecuador is approximately 1-to-2, which means that half of the country's expenditures on irrigation have been lost forever. Furthermore, attempts to make farmers pay for the recoverable remainder are stymied by their desire to avoid a decline in the value of real estate, which is the main asset of most rural households.

Under these circumstances, the prospects for full recovery of the capital costs of irrigation are discouraging in Ecuador.

The imperative of irrigation pricing reform

The impediments to improved cost-recovery in public irrigation projects are daunting, as recent experience in Ecuador shows. However, the issue of pricing reform is hard to avoid, for the simple reason that agriculture uses so much water.

Throughout the world, withdrawals for irrigation and livestock production from rivers, lakes, and streams as well as underground aquifers not only exceed extractions for other specific uses, but are also greater than combined withdrawals for industrial production and households' domestic use (see table 3.1).

The world's high-income nations are exceptional in that agriculture accounts for less than half of total extraction. This is a consequence both of the prevalence of rain-fed farming in temperate settings, such as the American Midwest, and the enormous volumes of water used to cool thermal generating plants and other industrial facilities.[5] In contrast, irrigated agriculture is widespread in Asia and North Africa. As a result, freshwater withdrawals for crop and livestock production exceed 80 percent of the total in these parts of the world. The same level is exceeded south of the Sahara (with South Africa and Botswana as exceptions), largely because there is not much industry and use of water by manufacturers is correspondingly limited.

In Latin America, along with Eastern Europe and Central Asia, freshwater withdrawals also exceed the combined total for other sectors. Needless to say, the proportion of water used in agriculture is small by regional standards in those nations with limited irrigation: 61 percent in Brazil, for example, where more than 95 percent

Table 3.1 **Freshwater withdrawals, by sector and region**

Region	Percentage used by:		
	Agriculture	Industry	Households
Latin America and Caribbean	74	9	18
East Asia and Pacific	81	14	5
South Asia	94	3	4
Sub-Saharan Africa	85	6	10
Middle East and North Africa	88	5	7
Eastern Europe and Central Asia	57	33	10
High-Income Nations	42	42	16
Entire World	71	20	10

Source: World Bank (2005), p.148.

of all farmland is rain-fed (World Bank 2005, 134, 146). In contrast, 84 percent of freshwater withdrawals in Chile – where many coastal areas receive little precipitation and 83 percent of agricultural land is irrigated – are for crop and livestock production (*ibid.*).

In light of the huge volumes of water used by farmers, the issue of irrigation pricing policy goes beyond recovering the cost of canals, pipes, pumping stations, and other infrastructure. Where water is scarce, as it is in most places, inefficient use of the resource in agriculture (which is induced by unrealistic prices) has repercussions for all other sectors. Especially where non-agricultural demand for water is increasing due to population growth, economic expansion, industrialization, and other factors, the opportunity costs of wasting and misallocating irrigation water become enormous. Indeed, it is no exaggeration to say that adequate supplies of clean water will never be provided to the population as a whole if there is no alteration of the pricing policies which enable farmers to neglect these opportunity costs entirely.

Pro-market water policies in Chile

Compared to the situation in most countries, Chile's water policies are unusually conducive to efficient resource use and development. Property rights are the salient feature of the Chilean regime. Since resource ownership is secure and transferable, markets guide the use of water, including its reallocation when and where appropriate.

In some respects, current Chilean policies are consistent with centuries-old traditions in the country and other parts of Latin America, where the notion of public dominion over water resources was inherited in the legal systems from the Spanish law. In turn, Spanish law inheirited this notion from Roman law, which did not consider continental waters as common resources but rather, considered them to be either public or private. Before independence, local associations of private farmers built, operated, and maintained irrigation canals. With subsequent legal development, however, the central government's role expanded. Chile's first Water Code, adopted in 1951, allowed state authorities to grant concessions to private parties. However, these concessions were transferable only if water was not used in any other activity, which essentially meant that all decisions about resource reallocation were in the hands of the state.

Governmental prerogatives were reinforced and extended in a revamped 1967 Water Code. Under this regime, all private rights were "administrative" – granted by the state for particular uses and entirely subject to public regulation. Moreover, use-rights were subject to expiration and water reallocation was determined by regional plans developed by the government. With the public sector exercising ultimate control over hydrological resources, uncompensated expropriation of water rights (and land) occurred during the latter part of Eduardo Frei's administration (1964 to 1970) and accelerated while Salvador Allende was President (1970 to 1973) (Bauer 1998).

Needless to say, the 1967 Water Code, which closely resembled contemporaneous legislation adopted elsewhere in the region (e.g., Ecuador's 1972 Water Law), was at odds with the free-market orientation that the military government adopted about a year and a half after the 1973 *coup d'etat*. Article 19 of the Political Constitution of 1980 was a clear repudiation of the Frei-Allende approach to water resource development. The Article established that "the rights of individuals over water, reserved or established in agreement with the law, will grant to their holders the property over them." This constitutional principle was put into practice in 1981, when Chile adopted a new water law.

The 1981 Water Code established that individual prerogatives in hydrological resources are property rights in every sense of the term, provided that ownership has been officially adjudicated by Chile's General Water Directorate (DGA). In addition to being permanent, water rights are transferable; sales are allowed either between farmers or between an irrigator and a non-agricultural user. Water rights, enforced by the DGA, can also be mortgaged, just as real estate can be. Furthermore, they cannot be expropriated without due compensation.

Significantly, the DGA cannot refuse to grant new rights if no one already owns the resource which is being claimed.[6] Hydroelectricity producers and other non-consumptive users are entitled to legal recognition and protection of their diversions from streams and rivers, provided that equal volumes are returned to the same channel. For consumptive uses, including irrigation, individual owners are entitled to withdraw a specific volume per time-period, although proportional reductions occur when stream-flow is unusually low (Hearne and Donoso 2005).

The 1981 Code mandated formal water rights for historical users – mainly irrigators (including small farmers who had benefited from the agrarian reforms of the Frei and Allende administrations), potable-water companies, and mines. Once this category of ownership was recognized, the DGA could create new rights in response to petitions submitted by resource users. The procedure governing the latter sort of adjudication begins with publication of proposed water rights in the *Diario Oficial* (i.e., the official journal of the country). If there are rival claimants, then the directorate organizes an auction, and the highest bidder takes ownership.

Throughout Chile, water users have responded to the 1981 Code by winning formal recognition of their historical rights and by acquiring resources which are not yet claimed by anyone else. But commercial water transactions, which the same law makes possible, have been less widespread. Such transactions are particularly rare where hydrological resources are abundant – in those parts of southern Chile with elevated precipitation, for example.

One part of the country with regular purchases and sales of water is the Limarí Valley, north of Santiago, where irrigated production of wine-grapes and other highly-valued crops has increased dramatically in recent years (Hearne and Easter 1997). From 1980 through 2000, nearly 28 percent of all water rights exchanged in the watershed were bought and sold independently of land transfers, with old and new owners making use of a market for permanent transactions created for exactly this purpose. In addition, a spot market exists in the Limarí Valley for resources used during a single growing season. In 1999–2000, for example, approximately 14 percent of the volume allotted to water users' associations in the region was exchanged in this market. During the unusually dry 1995–1996 season, this share was 21 percent (Cristi *et al.* 2000).

These transactions could not occur without the enabling legal framework created by the 1981 Water Code. In addition, water markets function well in the Limarí Valley because buyers and sellers have confidence in water users' associations, which maintain records of purchases and sales (*ibid.*). Another factor of great importance for the spot market is infrastructure, specifically the Paloma Irrigation System, which is the largest in Chile and the second largest in Latin America. Comprising three dams – Paloma (with a storage capacity of 750 million m^3), Cogotí (150 million m^3), and Recoleta (97 million m^3) – as well as an extensive network of canals, this system regularizes the supply of water and constitutes a guarantee to buyers that spot-market purchases will actually be delivered.

The organizational and institutional conditions that facilitate the spot market in the Limarí Valley are not fully satisfied in many other parts of Chile. As a result, exchanges of longer-term water rights have been the national norm, with these rights usually reallocated from irrigated farming to higher-valued non-agricultural uses. The Upper Mapocho Basin, near the Chilean capital Santiago, is a good example. In this area, sales to potable-water providers and real-estate developers accounted for 76 percent of the water rights traded from 1993 through 1999 (Donoso *et al*. 2001).

Notwithstanding the policy reforms which have made it possible for water resource use and development to be guided by market forces, new issues have arisen since promulgation of the 1981 Water Code. One of these issues is the monopolization of natural resources, especially in the hydroelectricity sector.

Taking advantage of its mountainous terrain and the abundance of water in many settings, Chile generates most of its electricity from hydro sources. Energy demand has risen substantially since the mid-1980s, as the national economy has expanded at a rapid pace. The 1981 Water Code encouraged hydroelectricity generation by creating property rights for non-consumptive uses of water, as discussed above. More than four-fifths of all such rights have been acquired by ENDESA, a Spanish firm that has invested heavily in the Chilean energy sector.

That ENDESA owns such a large portion of the resources that are suitable for hydroelectricity production raises obvious monopoly concerns. Similar concerns have been expressed about other categories of water rights, although the concentration of resource ownership resulting from the adjudication of historical ownership and new claims has been less extreme. At the urging of Chile's competition commission, the DGA has responded to ENDESA's dominance of hydroelectricity by refusing to grant new non-consumptive rights. This move has been supported by the Constitutional Court, which has ruled that the 1981 Code can be changed to allow additional conditions (including those meant to curb monopolization) to be placed on petitions for new water rights. Precisely this change was effected with new legislation in 2005.[7]

This same legislation created an additional tool for dealing with the concentration of hydrological resources in few hands. As of 1 January 2006, non-consumptive rights that are not being used – which according to the Ministry of Public Works exceed 80 percent of all such rights adjudicated by the DGA (Ministerio de Obras Públicas 2005) – will be subject to a fee. According to one consulting firm, ENDESA's payments for 2006 will be about US$2.6 million (Tanner Análisis 2005).

These recent modifications of the 1981 Water Code do not represent a repudiation of Chile's approach to water resource development in the past 25 years. With appropriate correction of the law to deal with excessive concentration of resource ownership, water rights remain secure and transferable, which means that allocation will continue to be guided by market forces. This market-driven system is highly advantageous in a country that has experienced rapid economic growth in recent decades, with direct consequences for water demand in agriculture, mining, and other part of the economy (Brown 1996; Figueroa *et al.* 1996). Indeed, it is doubtful that the competition over water resources inevitably created by economic expansion could have been resolved as effectively in the absence of policies that stress ownership and markets.

Summary and conclusions

Where water has grown scarce due to demographic and economic expansion, markets have proven to be an effective tool for resource allocation, including in developing nations. Markets require an enabling legal framework, which Chilean water law provides. In addition, the commercial exchange of water and resource rights is particularly active – both where demand is driven by highly-valued uses and where transactions costs have been lowered by institutional and infrastructural development, of the sort that has taken place in the Limarí Valley.

Other Latin American countries have attempted to replicate Chile's approach. Mexico's Water Law of 1992 resembles the 1981 Code by providing for the registration and transfer of water rights.

During the 1990s, Peru tried to adopt a water law that was consistent with the country's pro-market policies (Juravlev 2004).

But for the most part, the Chilean approach remains exceptional in the region. An objection frequently made by opponents of private water rights is that indigenous peoples are disenfranchised. However, this criticism is unjustified, since the combination of the 1981 Water Code and legislation subsequently adopted to protect such groups has given their rights precedence over other resource claims (Peña 2004).

Moreover, the failure to convert limited concessions for water use (of the sort that Chile had while the 1967 code was in force, and still exist in much of Latin America) into full-fledged water rights prevents everyone – including concession-holders – from capturing the gains created when markets are allowed to allocate resources to their most highly valued uses.

For example, an advantage enjoyed by those whose historical claims are formally adjudicated is that, simply by taking their rights to market, they can benefit from the efficient reallocation of hydrological resources. For example, irrigators who sell their rights to potable-water providers capture some of the value created when water is distributed to households rather than being applied to crops. But beyond this specific benefit, efficient markets for water and other factors of production – which can exist only if resource rights are permanent and transferable – are every bit as instrumental as trade liberalization in creating the right conditions for across-the-board economic growth.

As made clear in the case of Ecuador, subsidies and other forms of non-market allocation are not only financially unviable and economically inefficient, but are often injurious to the poor. By the same token, the case for water resource development driven by market forces is unassailable, as demonstrated amply by Chile's experience during the past 25 years.

Notes

1. Water use is either consumptive or non-consumptive. For example, the portion of water that is diverted from a stream, applied to farm fields, and then allowed to drain back into the same stream is classified as non-consumptive use, even though its quality may be diminished because it contains dissolved nutrients and suspended solids. The remainder of the diverted water, which evaporates or is absorbed by harvested crops, falls in the category of consumptive use.
2. The policy of selling potable water too cheaply sometimes created inefficiencies of an entirely different kind. For example, investments in infrastructure were not always constrained much by financial considerations, particularly when the national government paid for the difference between customers' payments and system costs. Under this circumstance, Quito's water company was known to use U.S. technical coefficients when deciding on the dimensions of pipes and machinery required for system extension. The capital expenditures resulting from this practice were excessive since per-household water consumption in the United States is several times per-household consumption in Ecuador (where extensive lawns, swimming pools, etc. are rare).
3. The resurgence of potable-water subsidies appears to be a consequence of populism. In recent elections, various mayoral candidates have promised cheap public utilities, including drinking water, for low-income districts. An upshot of this political competition is that 800,000 people in southern Quito no longer face any sort of volumetric tariff, but instead pay a uniform household-level fee of $5/month, which represents a very low level of cost-recovery.
4. On-farm water efficiency is defined as evapo-transpiration divided by total water delivered to the head of the main farm ditch.
5. As a rule, industrial cooling is a non-consumptive use of water. It usually is non-polluting as well.
6. The DGA can declare, however, that an aquifer is fully exploited and, on this basis, refuse to permit new withdrawals from the underground resource.
7. Specifically, the change was enacted with the passage of Law 20.017, on 11 May 2005.

References

Bauer, C. (1998). *Against the Current: Privatization, Water Markets, and the State in Chile*. Boston: Kluwer Academic.

Brown, E. (1996). "Disponibilidad de Recursos Hídricos en Chile en una Perspectiva de Largo Plazo," in O. Sunkel (ed.), *Sustentablidad Ambiental del Crecimiento Chileno*. Santiago: Universidad de Chile, pp. 191–213.

Carrión, R. (1993). "Evaluation of the Technical Assistance Provided by USAID/Ecuador and RHUDO/SA through the WASH Project to Quito's Municipal Water Company," U.S. Agency for International Development, Quito.

Cristi, O., S. Vicuña, L. de Azevedo, and A. Baltar (2000). "Mercado de Agua para Irrigación: Una Aplicación al Sistema Paloma de la Cuenca del Limarí, Chile," World Bank-Netherlands Water Partnership Program (BNWPP) Trust Fund, Washington.

Donoso, G., J. Montero, and S. Vicuña (2001). "Análisis de los Mercados de Derechos de Aprovechamiento de Agua en las Cuencas del Maipo y el Sistema Paloma en Chile: Efectos de la Variabilidad en la Oferta Hídrica y de los Costos de Transacción," XI Jornadas de Derechos de Aguas, Universidad de Zaragoza y Confederación Hidrográfica del Ebro, Zaragoza, España.

Figueroa, E, R. Alvarez, G. Donoso, J. Muñoz, and J. Lagos (1996). "Sustentabilidad Ambiental del Sector Exportador Chileno," in O. Sunkel (ed.), *Sustentablidad Ambiental del Crecimiento Chileno*. Santiago: Universidad de Chile, pp. 47–86.

Fujiwara, T. (2005). "A Privatização Beneficia os Pobres? Os Efeitos da Desestatização do Saneamento Básico na Mortalidade Infantil," Facultade de Economía, Universidade de São Paulo.

Galiani, S., P. Gertler, and E. Schargrodsky (2005). "Water for Life: The Impact of the Privatization of Water Services on Child Mortality," *Journal of Political Economy*, vol. 113, pp. 83–120.

Hearne, R. and K. Easter (1997). "The Economic and Financial Gains from Water Markets in Chile," *Agricultural Economics*, vol. 15, pp. 187–199.

Hearne, R. and G. Donoso (2005). "Water Institutional Reforms in Chile," *Water Policy*, vol. 7, pp. 53–69.

Juravlev, A. (2004). "Introducción," in *Mercados (de Derechos) de Agua: Experiencias y Propuestas en América del Sur* (Serie Recursos Naturales e Infraestructura N° 80), CEPAL, Santiago.

Ministerio de Obras Públicas, Transporte y Telecomunicaciones (2005). "MOP Celebra Junto a CONAMA Aprobación de Ley que Modifica Uso de Aguas." http://www.moptt.cl.

Peña, H. (2004). "Chile: 20 Años del Código de Aguas," in *Mercados (de Derechos) de Agua: Experiencias y Propuestas en América del Sur* (Serie Recursos Naturales e Infraestructura N° 80), CEPAL, Santiago.

Ricardo, D. (1965). *The Principles of Political Economy and Taxation*. London: Dent and Sons.

Southgate, D. and M. Whitaker (1994). *Economic Progress and the Environment: One Developing Country's Policy Crisis*. New York: Oxford University Press.

Tanner Análisis (2005), "Endesa pagará US$ 2,6 millones por derechos de agua no usados." 18 March. Online: http://www.elmostrador.cl/c_economia/tanner/t050318.pdf (visited February 14, 2006).

World Bank (2005). *World Development Indicators, 2005*. Washington, DC.

4 Poor provision of household water in India: How entrepreneurs respond to artificial scarcity

Laveesh Bhandari & Aarti Khare[1]

Though water is not strictly a 'public good', in most countries it has been a convention that water supply and provision is the government's realm. In India too this is the case.

India is blessed with some of the best natural water resources in the world. It has perennial rivers that are spread fairly evenly across the country, a large coastline, and (generally) high rainfall levels. The country's large population centers also tend to be spread out according to the availability of water. However, many urban Indian households do not have adequate water available for their daily requirements.

This chapter analyses problems with urban water supply in Delhi and other urban regions of India. (At least 700 million of India's people live in rural areas and similarly lack reliable and affordable water. This topic deserves consideration, but elsewhere.) It examines the negative human and environmental impacts of unpriced water which is supplied inefficiently (and in some cases not at all) by the public sector. We argue that good policy is about allowing and promoting private initiative, not preventing or controlling it. By suitably structuring the conditions for private sector involvement in water provision, governments would enable more

Poor provision of household water in India 93

widespread access to water across India, while simultaneously achieving environmental goals. The conditions for private sector water provision do not require micro-management or micro-regulation of the water supply economy, but would be facilitated by simple, broad policy measures.

Human water requirements

International organizations such as the U.S. Agency for International Development, the World Bank and the World Health Organization recommend between 20 and 40 liters of water per capita daily (LPCD) for the average human being. This estimate excludes water for cooking, bathing, and basic cleaning. These figures are similar to standards recommended by the UN's International Drinking Water Supply and Sanitation Decade and Agenda 21 of the Earth Summit.

Table 4.1 shows the estimates of per capita water requirements for a region with an average (moderate) climatic condition. However, for hot countries such as India, somewhat larger amounts are required – both for cleaning (better hygiene) and for consumption. The National Capital Territory of Delhi with its extreme climate provides a perfect example. The 2001 'Master Plan of Delhi' (MPD) recommends 70 gallons per capita daily (GPCD) (equivalent to 265 LPCD), while a manual on water supply and treatment produced by the Central Public Health Engineering and Environmental Organization's (CPHEEO) recommends 60 GPCD (227 LPCD) as a minimum.

According to these estimates, Delhi's daily water requirement in 2001–2002 was 827 or 965 million gallons (MGD), in contrast to the Delhi Jal Board's capacity to supply only 650 MGD. As discussed below, even this amount does not fully reach the household consumers.

India has adequate water resources to meet the daily needs of its inhabitants. There is also a water supply mechanism (by way of municipal supply or local water boards) which processes and supplies large quantities of water. However, our per capita water

Table 4.1 **Basic water requirements for human domestic needs**

Purpose	LPCD
Drinking Water*	5
Sanitation Services	20
Bathing	15
Food Preparation	10

Source: Gleik (2000).
*This is a true minimum to sustain life in moderate climatic conditions and average activity levels. In warm and hot climates the requirements would be somewhat higher.

requirement is extremely high – far higher than in most other countries. For instance, most European cities supply in the region of 120 to 130 LPCD. Even countries with a similar climate and per capita incomes as India – such as Senegal and Ivory Coast – supply in the region of 70 to 110 LPCD.[2] Most Indian cities do manage to process water at that level, and even beyond.

So where is the problem? Consider Delhi, India's capital: the city has relatively better water supply infrastructure than most Indian cities. Water is processed in the northern end of Delhi, from where it *flows* to the rest of the city. Unlike most of the world, urban water supply is not pressurized in much of South Asia, so water flows are a function of gravity and gradient. Though this system is less costly to operate, it also has serious environmental consequences.

Water requirements reach a peak during certain times, typically mornings. During this time, households draw the largest proportion of their daily requirements. But gravity-backed water flows are not adequate to service the needs during this peak time. Consequently, many households do not receive water and need to obtain supplementary water supplies (discussed below).

This problem is not limited to peak hour scarcity. Much of the water is not priced or is priced very low.[3] Municipal governments would actually receive negative net revenues if they established water-saving mechanisms (such as water meters), so little is done to encourage more efficient use of water. As a result, municipal governments have made few investments in improving water supply infrastructure. Theft and pilferage are rarely monitored and leakages are endemic. As a natural consequence, a large percentage of water is 'lost'. Estimates of these losses range from 40 to 60 per cent of the total water processed in Delhi. Almost all other South Asian cities face a similar situation.[4]

An associated problem is that of un-priced water in urban slums. These areas receive water at a common source, where it is collected by each household. In many places, this source is not even tapped, so water flows and drains away freely whenever it is supplied and regardless of whether it is collected.

While much of Delhi's water is wasted, most households do not receive adequate water and they resort to withdrawing sub-surface water. Many others find other ways to access it – such as illegally using pressurized pumps to draw water from the municipal pipes. These second-best solutions have many hidden costs:

- First, they involve large-scale withdrawals of sub-surface water. This causes unobserved and irreversible harm to the environment.
- Second, they are highly energy-inefficient, since many households have to draw water using their own motorized pumps. Typically, many large pumps drawing water require more energy than if water was adequately pressurized at a single point.
- Third, this requires households to store their own water. When water is priced artificially low or at zero, there is inadequate incentive for households to prevent waste.
- Fourth, poorer households purchase water from private or publicly subcontracted vendors. These vendors transport water tankers to neighborhoods which face water scarcity. This especially affects the poor, leading to a high unit cost of water, plus hidden costs in terms of additional effort and inconvenience.

Water supply in India differs from the rest of the world in another way: no Indian city supplies water for 24 hours a day. Almost all neighborhoods obtain water once or twice a day. Consequently they are forced to draw and store water for their *expected daily* requirements. During this drawing and storing, water is lost. Moreover, users are likely to draw more water than they actually need.

Leaking pipes, water storage and the slow movement of water during transmission and distribution contribute to health problems, especially for the poor. In India's tropical climate, parasites multiply rapidly. When water moves slowly, or is stored, this process accelerates. When pipes leak, impurities enter the water. Consequently, households are forced to filter and/or boil their water to make it fit for human consumption, a process that requires use of electricity and cooking gas (or LPG). Households that are unable to do so have negative health consequences – in the form of water-borne diseases such as diarrhea, typhoid and many others.

In the long term, better supply and usage of water will require it to be priced appropriately. Much water simply does not carry a price. Where prices do exist, in most Indian cities these are in the range of Rs. 0.5 to Rs. 5 (US $0.01–$0.12) per kilolitre.[5] Experts tend to believe that if water is priced at Rs. 15 (US $0.30) per kilolitre, even poor households have the ability to pay for water, and also have an incentive to use it more efficiently. At these levels, even local governments would have sufficient revenue to overhaul the water supply infrastructure and to implement water efficiency measures.

Many claim that the poor cannot pay for water, and use this argument to imply that water should not be priced. But this argument is not based on reality. India's relatively poorer population *does* pay for water, both in a monetary sense and in terms of the effort required to obtain clean water. Poorer households in a slum in Delhi pay Rs. 200 (US $4.50) per month for about 500 litres of piped water supplied twice a day (case study discussed in detail below). In a 30-day month, this works out to be Rs. 13.33 (US $0.30) per kilolitre. Add the convenience of 24-hour water, the cost of having storage devices and extra energy expenditures, and the price of Rs. 15 per kilolitre is well within reach of poor households. In any event, if the issue is increasing access for the poor, this is not an argument against pricing water per se, as there are many preferable alternatives to subsidize water. This, however, is not the main focus of this chapter.[6]

If household consumers can pay for water, it is possible to engage the private sector in supplying water. Private initiative in water supply has not been encouraged in the past (and even currently) in India. Environmental considerations have perhaps been the key factors behind opposition to the private sector. Consequently, many private water supply activities are illegal. In many cases even if these activities are allowed, government functionaries require bribery for the 'privilege' of allowing the activity to continue. Despite artificial barriers imposed to prevent private water supply, private sector initiative is growing. This is simply the result of demand which is unmet by public sector or government initiative, and likewise the speed and flexibility with which entrepreneurs can respond to changing conditions.

We show that private initiative, even when *not supported by public policy*, does not harm the environment any more than the government's inaction. In fact, private sector action may prevent existing problems from getting worse, and often leads to improvements in environmental quality. Perhaps more importantly, the private sector is able to service poorer neighborhoods much better than the public sector. Finally, when compared to their public sector counterparts, private sector entrepreneurs are far better at finding economically efficient ways to supply water in an affordable, efficient manner.

The public provision of water and existing scarcity

Water is under-supplied in India. The fact that water supply is almost wholly in the realm of the public sector verifies that governments are unable to provide adequate water supplies.

Table 4.2 Main sources of water in urban India

Source	No. of households (millions)	Percent
Tap	33.3	70.1
Tube-wells	10.2	21.4
Wells	3.2	6.7
Tank/ pond reserved for drinking	0.1	0.2
Other tanks/ ponds	0.0	0.1
River, canal, lake	0.1	0.2
Spring	0.0	0.1
Tanker	0.5	1.0
Others	0.1	0.2
Not available	0.1	0.1
Total	47.6	100.0

Numbers rounded to nearest decimal place.
Source: Data from NSSO (1999)[8] cited in Bajpai and Bhandari (2001).

Urban India is characterized by poor water supply infrastructure. Services are generally poor for all sectors of society, but for poorer sectors, conditions are worse because of their perceived inability to pay.[7] The government supplies water either by way of water boards or municipalities whose revenues are extremely low; a direct impact is a low level of investments and expenditures on urban services in general. Problems in water supply are especially severe.

This section reports the general conditions of water access by urban Indian households.

The figures in Table 4.2 pertain only to urban India. Approximately 70 percent of the urban households have tapped municipal water as their main source of water. The rest must rely on other sources.

Given that water processing centers must be located close to the main source of surface water, processed water travels a long distance before reaching many areas. Proper flow of available water is therefore crucial in determining water supply to the population. While the processing centres may have the capacity to process enough water to cater to the entire city, they may not be able to do so. Water is unavailable to many due to (1) lack of adequate supply infrastructure, (2) leakage and (3) illegal access from existing water lines during transportation. Some consumers break into the water pipe and draw water free of charge; the municipal employees (and sometimes the police) are paid regular bribes for allowing this to continue.

Consider the case of Delhi. In 1998–99, the total water processed and pumped by the Delhi Jal Board (DJB) was 2,475 MLD (million litres per day). Of this, at least 1,082 MLD – approximately 44 per cent of water supplied by the DJB – cannot be accounted for, which implies that the water is lost during distribution (Economic Survey of Delhi, 2001–2002). Water is lost mainly due to leakage in water mains, communication and service pipes and leaking valves (Suresh V. 1998). According to one estimate, about 82 percent of leakages occur in the house service connections, through service pipes and taps. The remaining 18 percent is due to leakage in the main pipelines. Moreover, water supply is un-metered in many urban areas. In lower-income neighborhoods, a significant proportion of water is supplied through stand posts, which also results in large, unaccounted for losses.[9]

Whether losses are due to leakage in ill-maintained pipes or due to pilferage by households and other entities, they impose two important costs. One cost is borne by the municipality, in terms of lower revenue; the other is borne by households that are forced to use alternative water sources to fulfill their daily water requirement or must make do with very small amounts of water.

Approximately 7 million households (20.5 percent) which receive municipal tap water must supplement their water supply with other sources (Table 4.3). Loss of water during transportation reduces the amount received by the households who are paying for the water that they use.

Given that water supply from the municipality is already priced at a very low level, and the cost of using alternative sources is very high, the tendency for households to supplement their water sources indicates that they receive an insufficient amount of water.

Table 4.3 **Supplementary water sources in urban India**

Supplementary source	Number of households with principal source of drinking water as tap water and supplementing	Percent distribution of households with principal source as tap and supplementing
Tap	614,564	1.8
Tube Wells/ hand pumps	3,992,515	12.0
Wells	1,521,794	4.6
Tank/ pond reserved for drinking	47,425	0.1
Other tanks/ ponds	40,074	0.1
River, canal, lake	209,466	0.6
Spring	68,572	0.2
Tanker	295,128	0.9
Others	79,876	0.2
No supplementary source	26,396,988	79.3
Missing	1,779	0.0
Total	33,268,180	100
No tapped water	14,172,772	

Source: NSSO (1999), authors' calculations

Approximately 4 million households (12 percent) which have municipal water as the main source use tube wells and hand pumps as a supplementary source. Of all the sources of water, installation of tube wells and hand pumps is the most expensive, requiring an initial lump sum expense (Rs. 100,000; US $22,727) for machine bore-wells and Rs. 90,000 (US $2,045) for hand boring (including pumps) in addition to maintenance expenses. Another 1.5 million households rely on wells to supplement municipal water supply.

The uncovered 30 percent

Thirty percent of India's urban population does not have access to municipal water and is forced to obtain water from other sources

(Table 4.2). Underground water – accessed through wells, tube wells and hand pumps – is the next most widely used principal source, accounting for more than 27 per cent of households' main water supply. A small percentage of urban households depend on water tankers rather than piped municipal water supply. Other sources such as tanks, ponds, springs, rivers, canals, etc. are also used, although insignificantly.

Good water supply

The true cost of water to the household depends on the ease with which it is accessed as well as the pricing system imposed by the water provider. Shared access and limited supply impose costs in the form of time spent in long queues. Access to water can be considered 'good' if processed tap water is available for 24 hours a day, is available inside the house and is intended for sole access. In contrast, many of the 70 per cent of 'tapped' households must share water from a main source. Of the 33 million households which do receive municipal water, only 15 million (46 percent) have exclusive access (Table 4.4), while 18 million (54 percent) require some sharing. Sharing essentially implies a single source – such as a public tap in a certain area – which is the main source of water for those living in the vicinity. Inevitably, this kind of access is characterized by queues and waiting, not to mention the inability to price such water.

As mentioned, the amount of water used by each household is measurable if the household has sole access. Not only does this

Table 4.4 **Tap water right of use in urban India**

Right of access	Number of households (millions)	Percent
Sole	15.2	45.6
Shared	8.7	26.1
Community	8.4	25.2
Others	1.0	3.1
Total	33.3	100

Source: NSSO (1999), authors' calculations

Table 4.5 **Relative distance from principal source of drinking water: Households with taps in urban India**

	No. of households (millions)	Percent
Within dwelling	15.3	45.8
Outside dwelling but within boundary (premises)	8.4	25.3
Others	9.6	28.8
Total	33.3	100.0

Source: NSSO (1999), authors' calculations

make it possible to charge for water use, it also facilitates differential pricing. If the responsible user can be traced, any overuse and misuse of water can also be easily identified and penalized. While punishment may help, incentives are a superior way to solve the problem amicably. A metered water supply would mean that more efficient use of water would be rewarded, since less use implies a lower charge. It also makes possible the management of water supplies at different times of the day – i.e. charging a higher price for peak times and a lower price for off-peak times. A metered water supply also makes detection of leakages is easier. Also, planning is easier if gaps in requirement and provision are precisely identified. However, shared access to water means that additional costs accrue, since it is difficult to measure usage by individual households and charge appropriately.

Much of the public discussion on water supply and user charges is based on the presumption that the households have *sole* access. However, this is not the case for the bulk of urban Indian households. A related problem is the distance between the dwelling and the source of water. That is, many households incur great effort to obtain water from the main source.

Only 46 percent of households have the luxury of having the tap within their dwelling; the rest must go outside their house to fetch water. Another 25 percent manage to have access within the boundary surrounding their dwelling. But 30 percent still must go

Table 4.6 **Water right of use in urban India**

Right of access	Number of households (millions)	Percent
Sole	19.6	41.3
Shared	12.7	26.6
Community	13.6	28.6
Others	1.6	3.5
Total	47.6	100.0

Source: NSSO (1999), authors' calculations

beyond the boundary of their dwelling. Even in urban India, almost 10 million households travel some distance to access tap water (Table 4.5).

Of approximately 20 million households who have sole access to tap water as their principal source (Table 4.6), only 11.3 million households have no supplementary source. We could conclude that these households are served the most efficiently. However they may lack a supplementary source of water, and as noted previously, 24-hour water is not supplied anywhere in the country.

In sum, not only is the penetration of municipal water supply low (about 70 percent of total households), it is also quite poor in terms of access. Most households that depend on tap water either share it with their neighbors, or transport it themselves to their dwelling, or both.

The poor in urban India

Of all the households that are not being serviced efficiently, the worst affected are those of fewer economic means. Many justifications for the prevalence of public provision of water are given, such as the idea that privatizing an 'essential good' like water could lead to high prices that would prevent the poor from having access to water. However, the existing public sector system imposes an additional burden on taxpayers, while the poor still do not have good access to tap water.

To assess the situation of the poor, a simple index was created to

Table 4.7 **Distribution of economic status as per principal source of water – urban India**

Source	Economic status of household			Missing	Total
	Low	Medium	High		
Tap	66	74	80	66	70
Others	34	26	20	34	30
Missing	0.2	0.0	0.1	0.1	0.1
Total	100	100	100	100	100

Source: NSSO (1999), authors' calculations

represent the economic status of the households. The index used ownership of certain amenities and certain lifestyle characteristics of the households (see Bajpai and Bhandari 2001). On the basis of this index, the households were classified as belonging to the 'low', 'medium' or 'high' economic strata. Approximately 41 percent of the total households belonged to the low economic stratum. Another 13 percent belonged to the medium economic stratum and 23 percent to the high economic stratum. The remaining 23 percent could not be characterized into these categories since not enough information was available on their asset ownership/use (Table 4.7).

Looking across various economic sectors of society, Table 4.7 shows that a high percentage of households receive tap water. Looking across the figures for all households, the distribution between tapped and untapped is 70:30. In case of the low economic class, the ratio is 66:35. In the case of households with medium and high economic status, the ratio is 74:26 and 80:20 respectively.

The absolute numbers are more alarming than the distribution suggests. Approximately 7 million households (14 percent) in urban India that belong to the lower economic stratum do not have access to tap water. Public provision of water has not managed to make tap water accessible to the urban poor. The public sector hasn't even managed to provide good water supply to relatively wealthier households which do possess the ability to pay for tap water.

Environmental impact of poor water supply: Overuse of sub-surface water[10]

Of 47.6 million urban households, 20 million extract water from underground sources. Of these, 13.4 million rely on sub-surface water as their principal source. The pressure on groundwater use is not reduced by public provision of water. Even among those who use municipal tap water as their main source, there are many who supplement it with other sources. A direct impact of the insufficient public sector provision of water is that wealthier people can afford to access alternative sources. In urban areas, this tends to be sub-surface water resources.

Groundwater can be an efficient source of water supply only if it is not *over-extracted*. As is well known, groundwater is over-extracted if it is extracted in excess of the groundwater recharge. Groundwater replenishment (recharge) stems from rainwater infiltration. An underground aquifer can become extinct if water is extracted at a faster rate than the recharge rate.

Over-exploitation of groundwater has the following environmental effects which may not be taken into account when establishing the economic cost of depletion:

◆ It causes water tables to recede to such low levels that the aquifer cannot replenish naturally. This causes the source of water to become extinct over time.
◆ Pollution of fresh water increases. Groundwater is available at shallow as well as deep levels. Deep groundwater constitutes the fresh water system. Recharge from rainwater results in replenishment of the shallow aquifer by means of upward leakage from the deeper aquifer. The decline in groundwater levels due to overexploitation changes the hydraulic gradient, thereby triggering the speedy movement of pollutants from above to the deeper groundwater system.
◆ Land subsidence (the lowering of the land surface) results from changes which occur underground. Subsidence is generally permanent, because even if the aquifer is recharged

to original levels, this would not cause the land surface to elevate (Leake 1997). Land subsidence causes many problems:
(1) Slope of streams, canals and drains become altered.
(2) Bridges, roads, railroads, storm drains, sanitary sewers, canals, and levees as well as public and private buildings may be damaged.
(3) Forces generated by compaction of fine-grained materials in aquifer systems may cause failure of well casings.
(4) In some coastal areas, subsidence may cause tides to move into low-lying areas that were previously above high-tide levels.
(5) Earth fissures are also associated with land subsidence. When groundwater is pumped, it causes horizontal movement in the sediments. This leads to narrow cracks of an inch or less in width. Over time, these cracks can expand and develop into large fissures, tens of feet in width. This is due to erosion, as the fissures intercept surface drainage.

Groundwater is a mineral resource and has a dynamic character, with both stock and flow aspects. If only the flow of the resource is utilized, then it is possible to sustain the use of groundwater over time. If, however, we also utilize the stock, then the resource tends to diminish over time, causing environmental damage along the way. Generally if the subsurface water table is falling, it indicates over-extraction of groundwater.

A constant monitoring and control of groundwater utilization is required to ensure environmentally sustainable use of groundwater. This requires knowledge of the situation regarding the existing groundwater resources. However adequate assessment of groundwater use for non-irrigation purposes is still lacking (Dhawan 1995).

As a reply to a question raised in the Indian Parliament in 2000 (Lok Sabha 2000), a State-wide list of over-exploited and dark blocks was obtained from the Ministry of Water Resources. A dark block is on the verge of being over-exploited; over-exploitation is defined as extraction in excess of recharge. According to these figures, of the

5711 blocks, *taluks*, *mandals* (administrative units) and watersheds, 310 were categorized as dark and 160 were marked as overexploited.

In India, the largest environmental impact of receding groundwater levels is the increased scarcity of potable water. Evidence specifically pertaining to urban areas is not well documented. However, the impact of over-extraction of groundwater is obvious from various rural areas of India where groundwater is drawn by hand-pumps and recently by means of tube-wells.

The Center for Science and Environment has also noted that over-pumping, lack of groundwater recharge and a gradual destruction of the local traditional systems of water harvesting have contributed to water shortages (CSE 1999).

For instance, in Chennai the overexploitation of groundwater caused water tables to fall, and regions around the sea experienced ingress of seawater which then led to extreme soil salinity. The same is true of Junagarh district in Gujarat, and many other areas of India.

Over the years, groundwater contamination has caused many deaths in India. The fact remains that most areas which currently have contaminated water previously depended exclusively on groundwater. As these areas were already rich in certain metals or compounds, falling groundwater levels meant increasing contamination. In Madhya Pradesh, the district of Mandla is one example where consumption of groundwater leads to Fluorosis. Approximately 94 per cent of people in the 157 villages in the Dungarpur district of Rajasthan suffer from dental Fluorosis, and an abnormally high 32.5 per cent have skeletal Fluorosis, which affects about 100,000 people in the state.

Arsenic pollution has also been noted, particularly in West Bengal where groundwater in many of the State's small towns and villages is now contaminated. When arsenic is naturally present, overuse tends to concentrate its levels. In Bichhri, a village in Rajasthan, the water pouring out of the bore-wells is brown in colour. More than 90 wells which were once used for irrigation and domestic water now lie unused (CSE 1999).

Table 4.8 **Delhi's groundwater decline during the 1990s**

Blocks in Delhi	Decline (meters)
Mehrauli	4–10
City	4–8
Njafgarh	4–7
Kanjhawala	4–5
Alipur	4–5

Source: Lok Sabha (2000)

Groundwater use has increased as urban India's population grows and public water provision becomes relatively more scarce, and the effects are beginning to show. Long-term observations about Delhi's water supply made by the Central Ground Water Board (CGWB) have shown that, in many parts of the city, groundwater levels declined drastically during the 1990s.

The maximum range of decline is 4 to 10 meters in Delhi during the last ten years. The situation is similar for other cities that do not have good access to processed tap water. The inability of the local governments to provide water (a duty of the local government by law) and the need for more water has led to the emergence of alternative means to obtain water, which generally utilize groundwater. The inability of the governments to enforce poorly-defined groundwater laws is visible from the emergence of private water suppliers.

Private sector responses

Poor provision of water by the public sector has meant that low quality water has been delivered in an extremely inefficient manner. This means poor coverage of households and the denial of a basic necessity, creating an unmet demand for water.

To get around problems caused by the public sector, wealthier urban residents have constructed their own tube wells – an activity that is not strictly illegal. However, the poor are unable to afford this alternative. Part of their water requirements are being met by informal entrepreneurs operating in the private sector. This activity occurs in two ways:

1. Tankers operated by private individuals and companies supply the water either from surface water sources (legal) or from subsurface sources (illegal). Water is priced according to the volume drawn from each water tanker; and
2. Water is supplied from sub-surface sources and transported by way of pipes to within the household premises. This activity is considered illegal by the government.

Private water tankers

The Delhi Jal Board (DJB) is the government's water supply arm. The DJB has its own fleet of tankers for water distribution. However, these tankers are only used in the following cases:

- When water is not available because of leakages or bursts in water lines, or any other faults in the system. In such a case the water tanker is to be supplied within three hours of complaint, subject to availability. This service is provided at no cost to the users.
- For private functions such as marriages and religious functions. The water tanker can be supplied on any working day, if a booking is made 15 days in advance. The price depends on the distance of the house from the DJB water storage area (Table 9).

It is not possible to buy water from the DJB on short notice; an advance booking is essential. This leaves many people, who may or may not have a municipal water connection, without any recourse from the government when they are stranded with no water supply.

To solve temporary water shortages, communities or households may request the services of private tankers. Private tankers are contracted by municipalities to supply water. This water is accessed either from a surface or sub-surface water source, and this involves little (if any) processing. However, there is no difference in the price

Table 4.9 **Tanker rates from Delhi Jal Board**

Distance	Stationary (Rs./US $)	Filling (Rs./US $)
Up to 5 kms.	Rs.400/$6.82	Rs.225/$5.11
5–10 kms.	Rs.600/$13.64	Rs.325/$7.39
Beyond 10 kms.	Rs.1000/$22.73	
Upto 15 kms.	Rs.425/$9.66	
Beyond 15 kms.	Rs.450/$10.23	

Source: Delhi Jal Board (2002).
Approximate prices based on exchange rate of Rs.48/US $1.

level for the different sources of water. Some private suppliers supply processed (DJB) water; they are subcontracted by the DJB to do so.

Private water tankers transport water and fill it into the household's storage tank. They charge a standard fee according to the quantity of water supplied, and do not require an advance booking. The water tankers generally have a capacity of 3,000 or 6,000 liters. The charge is Rs. 300 (US $7) for 3,000 liters and Rs. 500 (US $11) for 6,000 liters. Since these private suppliers are present in almost all neighborhoods, the prices generally do not relate to the distance the tanker must travel.

Often, groups of people in the same neighborhood collectively call for a tanker. These are generally low-income or lower-middle class households which may not be able to accommodate 6,000 or even 3,000 liters (either because their storage tanks are smaller, or they don't possess a storage tank). These households share the cost of water by paying for the quantity that they individually consume.

Private piped water supply: A case study from Delhi

The case study is from an illegal (squatter) settlement (slum or *basti*) situated in South Delhi which faces acute water shortages, especially in the hot and parched summer months. All four of Delhi's municipal water processing centers are located in the northern part of the city; by the time the water reaches the southern part of the city, much of it is lost.

The residents of this *basti* and many similar *bastis* in Delhi form a major portion of the labor force for the city's service industry. While the government is 'charitable' enough to provide water by means of occasional tankers, this water is not provided to them as a right or even as a service for which they would be willing to pay.

The DJB's supply network exists in most high- and middle-income neighborhoods. Though these households also suffer from water shortages, the situation of the residents of illegal settlements is much worse. In the *basti* studied by the authors, the DJB does not provide piped water.[11] These areas have hand-pumps, installed by the DJB, which draw sub-surface water. Most households in this *basti* rely on these hand-pumps for their daily water needs.

The number of hand-pumps compared to the number of users is very small, leading to long queues. Users must carry water in many vessels simultaneously to avoid queuing too often. Water is then stored in small vessels in the house and households are able to use it only minimally. The deep groundwater aquifer is the source of fresh water. In summer months, when the water tables fall, the hand pumps do not provide adequate water, and since the hand pump reaches a relatively shallow depth in the ground, the water is also not very pure. When the situation becomes dire during the summer, residents approach the local politicians, who then send water tankers from the DJB to the *basti*.

In order to avoid the trouble and the uncertainty associated with accessing water from the hand-pumps, many residents of this *basti* now buy water from private sector suppliers of piped water. This has provided an adequate solution for many households, who are willing and able to pay for their water.

In one area of this *basti*, piped water is supplied to approximately 80 to 90 households by a private firm. This water is obtained from the ground (at a depth of around 150–160 feet) by means of a pump. The boring for extraction of water was done mechanically, at a cost of approximately Rs. 40,000 (US $909)– a cost which no household in the area could bear on its own.[12] A pump was acquired for Rs.

25,000 (US $568), so the total starting cost was therefore about Rs. 65,000 (US $1,477).

The households were required to pay Rs. 500 (US $11) as a security deposit for access to this service. The terms of the agreement are simple: if the household wishes to discontinue taking water from the supplier, then their security deposit is not refunded. On the other hand, if the supplier decides to discontinue supply then he must return the deposit to the households. Many households in this *basti* are not connected by this water supply because they are unable to pay a lump sum of Rs. 500.

In addition to the deposit, the households paid the cost of the initial supply network of pipes, which amounted to another Rs. 500. The households were free to get the pipes fitted on their own. The arrangement for the maintenance of the pipes is also quite simple: The main pipe that runs through the road connecting the household is the responsibility of the suppliers; from the main pipe to the house, the pipe is the responsibility of the respective household.

The firm agrees to supply water for half an hour, twice a day, with good pressure, which is enough to fill a storage tank with a capacity of 500 liters. The tank costs approximately Rs. 1000 to 2000 (US $23 to $45). In case the tank does not fill up during this time, then the suppliers continue for a little longer on request. There is no way to measure the amount of water supplied, as neither the suppliers nor the households have a meter installed.

Seven to eight households receive water at one time. At the time water is being supplied, the owners of the water supply are watchful of pilferers.

For this service, the households make an advance payment of Rs. 200 (US $4.50) per month. However, the terms of payment are not very severe. The private suppliers – who live in the same neighborhood – typically accept a delay of a month depending on the household's economic circumstances.

The supplier's cost of providing this water supply is mainly in running the pump. Though an electricity connection is available, the pump uses a diesel generator. The pump runs for 7 hours daily, at a cost of approximately Rs. 9000 (US $204) per month.[13] Since water is supplied to 83 households, the supplier generates a surplus of Rs. 6,400 ((US $145) per month. In addition, on certain occasions, other households buy water from the supplier at a rate of Rs. 30 for 100 liters. These households generally rely on the hand-pumps in the *basti*, but when they require more water for some special event, they choose to buy it from these suppliers.

Private water supply systems like this one connect approximately 40 percent of the households in this *basti*. All the water suppliers in this area supply water more or less at the same price.

Prior to the existence of this private supplier, the households spent significant time and effort to obtain small amounts of water from the hand pumps. In summer when the hand-pumps would dry up, then they would have to go to hand pumps in other areas that were further away. The last resort was a plea to the local politician, who would then ask for a DJB tanker to be sent to the *basti*. An additional hassle faced by these households was fights that broke out early in the morning when people left for work. To avoid all of these costs, these households have chosen to pay Rs. 200 per month – approximately 6.6 per cent of the average monthly salary in the neighborhood – to have water piped directly into their homes. They receive higher quality water since the water is drawn from deeper into the ground than the hand-pump.

Discussion

Significantly, in this system of informal water supply prices, it is access to water (rather than the water itself) which is priced. The monthly charge of Rs. 200 includes access to as much water as the household can draw during the half-hour periods in which water is made available. Household consumption is also limited by the limited capacity of the pipes.

It is well known that the most efficient system of water pricing is based on marginal cost principles (that is, a per unit basis). With marginal cost pricing, delivery systems have the right incentives for producers to supply optimal levels, for consumers to use water efficiently, and as a consequence, the process has a minimal negative impact on the environment. However, the informal system of priced access and unpriced supply in this *basti* is somewhat removed from the marginal cost principle.

The key question, of course, is why this method of pricing has not emerged. Economic principles might indicate that a completely free system would tend to move towards a pricing system based on marginal cost. Are there other forces, technical or otherwise, which affect this market?

Two issues are related to marginal cost in this market: first, the cost of extra per unit of water and second, the extra cost per unit of time. We argue that the extra cost per unit is zero within a certain range, but the extra cost per unit of time is significant. Consequently, it is in the supplier's interest to price access (based on a unit of time) and not to price units of water.

First, the supplier must extract water and supply it at pre-determined minimum pressure levels – otherwise it does not reach households which are more distant, located on higher floors, or at a higher elevation. The diameter of the pipe is also fixed. Consequently, water must be supplied at the given pressure levels, or not at all.

Second, and corollary to the former, sub-surface water extraction cannot occur below a minimum pressure level, and after it is extracted, it is costly to store. If the water was stored, the supplier would not only incur storage costs, but also the costs of re-pressurizing the water. The most energy- (and cost-) efficient system for this market is one where water is extracted and supplied simultaneously, using the same motor.

Third, pricing on the basis of units of water requires the installation of water meters in each household. These are costly, and are not tamper-proof.

This informal market is not the most efficient in terms of environmental considerations, which would require water to be priced on a per-unit basis. By paying the price of access alone, users have some incentive to draw and use as much water as possible. There is no counterbalancing force for misuse or inefficient use. However,

Figure 4.1 **Scale of environmental efficiency versus pricing**

Free water for all   Pricing limited to access   Per unit pricing

---

this system is a great improvement on that which is effectively instituted by public sector/government provision – where poor quality water is supplied and priced inadequately.

Moreover, environmental considerations are not the only objective of public policy towards the water sector. Ample coverage of households is equally – if not more – important. Yet the government is seemingly unaware that water can be supplied by entrepreneurs in the private sector to those households which the government neglects.

Delhi's current laws, for instance, allow households to access surface water for their own individual requirements. Likewise, the laws allow similar access to businesses, which then manufacture bottled drinking water. Both of these activities are effectively only for wealthier consumers, who can afford to spend large amounts for their private water needs. They are also the primary consumers of bottled drinking water. Poorer sectors of society can only access subsurface water through an entity that can supply them with water on a commercial basis – but that option is prohibited by Delhi's laws.

Such regulations and controls are intended to reduce environmental impacts (their effectiveness is another matter altogether). Yet they have perverse consequences for the poor. A good policy towards water provision has to create a suitable middle ground, a level playing field for water to be supplied in various ways. These issues are discussed in the next section.

Policy implications

India's municipal governments have been unable to price water adequately, and this has resulted in their inability to finance service and water supply infrastructure improvements. It has created an artificial water scarcity for many households, and has also encouraged misuse and waste of water.

Current efforts to reform the sector will be limited by the poor quality of current services and infrastructure, the perception that the poor cannot or should not pay for water, and the inability of government functionaries to recover revenue, even when water is priced. Each of these factors contributes to the inability of the government bodies to provide adequate water for households.

However, without per unit charges imposed on water it is likely that water will be misused or wasted. Moreover, without the ability to generate funds from within the sector, cash-strapped local and state governments will be unable to improve existing infrastructure and services, much less expand them to meet the increasing requirements of a growing urban population.

This catch-22 situation exists in almost all cities and towns in India. In a very few of the larger cities, external agencies have stepped in to provide finance concessions and advice – but there are more than 000 cities and towns.

Any improvements in infrastructure and service require capital, and this capital can be accessed from commercial sources. To do that, a revenue stream from water charges would be essential. Indeed, this revenue stream is possible, as we show in the next subsection.

Breaking the vicious cycle: Good water policy

Water supply can be divided into three broad stages: Processing, transmission, and distribution (Figure 4.2). Currently, all three stages are the responsibility of government. A pricing system can only be established if input and output can be monitored. When monitoring is possible, either the government or the private sector can directly undertake these activities. Financing and pricing then

Figure 4.2 **Broad stages of water supply**

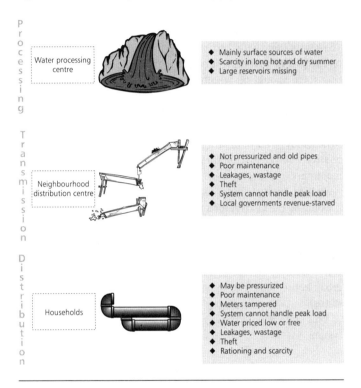

becomes much easier. Simple subsidy mechanisms also then become possible.

We start with the last stage – distribution. This is because where revenues may be recovered and if the water supply mechanism operates smoothly at this stage, then the rest will follow. However we do not make specific policy recommendations, instead we suggest a broad policy framework. The specific policy aspects are highly dependent on local factors, and should only be developed on a case-by-case basis.

The distribution stage

The key problem with distribution is the inability and unwillingness on the part of government bodies to recover costs. For that reason, this stage should be wholly in private hands: the water can be supplied to distributors at a price, for which the distributor can charge a mark-up to the consumer. The amount of water supplied to the distributor can easily be measured by the transmission agent (which may or may not be a government entity); and since there are relatively fewer distributors than consumers, it is far simpler to monitor.

In turn, the distributor would supply water to the consumer and it would be the distributor's responsibility to recover revenues from consumers. The government could, of course, subsidize the cost of water meters in the poorer areas. Water for the poor could also be subsidized, and there are many ways of so doing. Some examples are water stamps, direct subsidies, or area-specific water rates (since urban households tend to be clustered according to their economic status).

This system would have numerous advantages over the existing system.

1. The costs of any water losses in the distribution stage are automatically borne by the distributor, and it has a strong incentive to prevent waste or theft (as observed in the case of the private water supplier discussed above).
2. Water is not given for free; those who waste the water are required to pay for it.
3. The consumer pays the distributor, who thereby has an incentive to collect the charges from the consumer as its returns depend upon the revenues it receives.
4. Moreover, as discussed in the case study above, a private distributor has an incentive to show some flexibility where

payment is concerned. In contrast, a bureaucratic public sector system rarely has such flexibility; revenue collection often encourages bribery and corruption on the part of government functionaries.
5. Subsidies can be directed to those who are most in need. More importantly, subsidies need not be unlimited.
6. Private entrepreneurs have an incentive to reduce costs, take risks and engage in innovative behaviors, in contrast to public sector distributors which are run by salaried government employees.
7. Since the firm charges a per-unit mark-up to the consumer, it is in the firm's interest to supply adequate quantities of water, but not to waste it. Similarly, it is in the interest of the households to not waste water.
8. Shifting to this system would require no large-scale, up-front investments; distribution of water could be privatized in the system that currently exists.
9. The private distributor has an incentive to provide 24-hour water at the appropriate pressure level, and it is therefore in the distributor's interest to invest in quality and efficiency improvements.
10. Depending upon the particular system instituted by the government, it may not even be necessary to carry out detailed account-keeping and regulatory activities. Transactions can occur merely on the basis of amount of water sold. This implies that the distribution stage need not have formal private sector distribution companies. Informal (unorganized) private entrepreneurs can also be allowed to resell water, with the benefit that these individuals are far better placed to act on local knowledge. In turn, this implies that we do not even require local water distribution monopolies; a market for private distribution of water can develop.

In general, the overall benefit of disassociating water distribution from transmission and production is that it would create flexibility. Different systems could be followed depending upon the local conditions.

Water transmission

The transmission stage essentially requires a transmission agent, which may or may not be a government entity. The entity's primary task would be to prevent water from leaking and being stolen. In the long term, as Indian cities also shift to pressurized water supply, almost all the infrastructure at this stage needs to be overhauled. Since the input (from the processing stage) and the output (sold to the distributor) are both measurable commodities, the problem of transmission is simple in economic terms. However, the process of improving the system, preventing leakages and theft, and overhauling the infrastructure is operationally a very complex one.

The key issue, however, relates to adequate revenue generation. It is not clear whether revenues alone would enable the capital requirements for the large scale overhaul which is needed. In case they are not, some form of government subsidy could be needed; this would have to be judged on a case-by-case basis. In any event, a monitored system that generates revenues would be an essential first step for their overhaul.

Water processing

An urban water supply system depends on an adequate source of water. This is broadly possible, since most of India's cities are located close to perennial rivers or large lakes. However, in the long summer months, the water levels in many rivers reduce and the water levels in reservoirs or lakes are substantially reduced (and sometimes even dry up). This requires an extension of the lakes, tanks, or reservoirs. Minor dams on rivers that trap the surplus water during non-summer months would also help. All of these require capital. Water processing plants in many cases are quite old and need to be improved and their capacity expanded. This also requires some investment.

As in the case of the transmission stage, the processing stage can also occur in the private sector; there is nothing preventing this.

Regulation and markets

It is generally considered that if the private sector were to provide water supply services, some regulation would be essential to prevent monopoly exploitation. However, this is not necessary. The private sector has the ability to provide quality services at a low price provided that sellers are competing with each other. For very large cities in India, it is possible to have more than one or two entities which process and transmit water. Where distribution is concerned, a great deal of competition would be possible in most cities. Private water supply need not be equated with regulation. At the same time, where only one or two companies are feasible, some form of price and quality regulation may be the only way to enable the private sector to play a role.

Conclusion

India's government water suppliers have created an artificial water scarcity in urban areas. At the same time, municipal governments are oblivious to how this problem is being addressed by the private sector (and, remarkably, they do so despite numerous government barriers).

Given the right conditions, entrepreneurs acting in a private capacity (whether formal or informal) have an important role to play in delivering high quality water to households. Where environmental considerations are concerned, it is clear that pricing water leads to less waste and more efficient use, with related environmental benefits. Moreover, private sector involvement may prevent existing problems from getting worse, and often leads to improvements in environmental quality. Given enough freedom to operate and a supportive legal framework, private sector entrepreneurs would also have a positive effect on the existing inequality of service provision and coverage for the poor.

Where processing, transmission and distribution of water is concerned, it is possible to have a unit-based pricing system for water, even in the existing system. Unit-based pricing would create more efficient use of water, and would enable all households to be supplied with high-quality water. However, given conditions in urban India, this is only possible if water supply is left to the private sector at the distribution stage.

This chapter does not intend to present a detailed analysis of the economies of scale, or financial implications of such a system. (Although this would be essential before such a system were established.) It simply demonstrates that such systems are possible, and not too difficult to establish.

Notes

1. We would like to thank Peeyush Bajpai, Mridusmita Bordoloi, Amar Gujral, and Dhyan Singh for their useful comments and invaluable support. We also referred to Anderson and Snyder (1997), de Villiers (1999) and Holden and Thobani (1996) for the general ideas and principles which guided this paper. However, all remaining errors are our own. Comments appreciated at laveesh@indicus.net
2. We would like to thank Dr. Vivek Srivastava of the Water and Sanitation Program, The World Bank, New Delhi, for this information.
3. Generally the reasons ascribed to this are related to inability of the poor to pay. That the water service is poor also contributes to the inability of the government to charge higher prices. But low or no pricing also creates another problem.
4. For instance Katmandu, Nepal, which recently shifted to a pressurized system. The city faced even higher rate of loss, reportedly in the range of 70 per cent.
5. Rupee to dollar conversions assume an exchange rate of Rs. 44 to one US dollar (prevailing exchange rate in February 2005). Small price figures are provided to give the reader an idea of relative prices.
6. For a discussion on these issues see World Bank (2002).
7. As in other poor countries, a large part of urban India lives in slum-like areas where municipal services are almost non-existent. Moreover, this situation shows no signs of improving due to the

generally poor financial situation of urban local governments across India; also see Srivastava and Sen (1997) and Khandwalla (1999).
8. In the NSSO 1998 survey, 110,313 rural and urban households were sampled in a representative manner. The survey is used as the basis for subsequent tables in this chapter.
9. Stand posts are stand alone water pipes that are located in low income areas. Many are untapped. Even when tapped, they are rarely maintained properly. And as a result large amount of water is wasted.
10. See Leake (1997) for a detailed discussion.
11. Recently, the DJB has given the people in unauthorized colonies an option to access municipal water:
    UNAUTHORISED UNREGULARISED COLONIES
    - Rates for water development charges for provision of water lines in unauthorized unregularised colonies has been revised from the existing Rs. 55/- per sq. meter to Rs. 110/- per sq. meter. If the payment of full amount of development charges is made within three months of the receipt of first bill, 10% rebate will be given.
    - Water lines will be laid on payment of first advance installment of Rs. 25/- per sq. meter of the plotted area.
    - The tender will be called only on receipt of 50% of the first installment and work will be executed only after receipt of 75% of the first advance installment from the colony [neighborhood] as a whole.

    After execution of the scheme water connections will be released on payment of another Rs. 25/- per Sq. Mt. and the balance will be payable in 20 equated quarterly installments. (Source: Delhi Jal Board) Any such arrangement would require all the residents to contribute otherwise it will not materialize. Once the infrastructure is set into place it might be a cheaper option, but the efficiency may still be questionable.
12. Hand boring is a slightly cheaper option, but after digging to a certain depth it is impossible to dig any lower without the use of machines, since there is a layer of rock.
13. An amount of money (not revealed) must also be given to the police.

References

Anderson, Terry L. and Pamela Snyder (1997). *Water Markets: Priming the Invisible Pump.* Washington, D.C.: Cato Institute.

Bajpai, Peeyush and Laveesh Bhandari (2001). "Ensuring Access to Water in Urban Households." *Economic and Political Weekly*, September 29, 3774–3778.

Center for Science and Environment (1999). "Perpetual Thirst: Faucets of the problem", *Down to Earth* Vol. 7, No. 19, February 28. Website: www.cseindia.org

de Villiers, Marq (1999). *Water: The Fate of Our Most Precious Resource.* Canada: Stoddart Publishing Co. Limited.

Delhi Jal Board (2002). www.delhijalboard.com/w_con.htm#dc (Delhi Jal Board)

Dhawan, B.D. (1995). *Groundwater Depletion, Land Degradation and Irrigated Agriculture in India*. New Delhi: Commonwealth Publishers.

Economic Survey of Delhi (2001–2002). Website: http://delhiplanning.nic.in/Economic%20Survey/Ecosur2001-02/Ecosur2001-02.htm

Gleick, Peter H. (2000). *The World's Water: The Biennial Report on Fresh Water Resources, 2000–2001*. Washington, D.C.: Island Press.

Holden, Paul and Mateen Thobani (1996). "Tradable Water Rights: A Property Rights Approach to Resolving Water Shortages and Promoting Investment." World Bank Policy Research Working Paper No. 1627 (July).

Leake, S. A. (1997). "Land Subsidence from Ground-Water Pumping". Paper presented at Impact of Climate Change and Land Use in the Southwestern United States (conference). U.S. Global Change Research Program. U.S. Geological Survey.

Lok Sabha (Indian Parliament) (2000). Unstarred Question No. 2792. Website: http://164.100.24.208/lsq/quest.asp?qref=11578

NSSO (National Sample Survey Organisation)(1999). 54th Round, Water and Sanitation Report, Report No. 449. Government of India, New Delhi. Website: http://mospi.nic.in/mospi_nsso_rept_pubn.htm

Suresh, V. (1998). "Strategies for Sustainable Water Supply for All: Indian Experience." Working Paper presented at "Water and Sustainable Development" (conference). International Office for Water, Paris.

World Bank (2002). "New Designs for Water and Sanitation Transactions: Making Private Sector Participation Work for the Poor" Water and Sanitation Program. Washington, DC, May. Website: http://www.wsp.org/publications/global_newdesigns.pdf

5 The rain catchers of Saurashtra, Gujarat

Ambrish Mehta

India's western state of Gujarat recently has faced a great deal of turbulence in its water sector. Name a water-related problem, and one or another part of the state is afflicted with it. These problems range from chronic water scarcity and droughts to floods and water logging, mining of deep underground aquifers to water logging and secondary salinity in canal-irrigated areas, intrusion of sea water in coastal areas, unviable public water service systems for urban and rural water supply and canal irrigation projects, discharge of untreated sewage and industrial effluents in the rivers, and pollution of drinking water with fluoride, nitrates and other pollutants.

However, Tushaar Shah of the International Water Management Institute (IWMI) makes the astute observation that "There are few regions in the world where water has become everybody's business quite like it has in Gujarat, especially in water-scarce regions of Saurashtra and Kutch" (Shah 2002).

For over a decade, a broad-based mass movement for capturing rain water to recharge ground water has occurred in these two regions of the state.[1] Today, it is still going strong and showing no signs of fatigue. Individual farmers, village communities, local voluntary organizations, activists, non-governmental and community-based organizations, religious and spiritual organizations, Gram Panchayats (local government councils) and municipalities all are catching rain water. They store it in wells, ponds, tanks, check dams, drains, or wherever they can, and find ways to put it under the ground so that it can be retrieved and used later. Across the entire state, there is broad-based support for these "rain catchers" of Saurashtra.

What is striking is that this movement is entirely local and homegrown. Local philanthropies and voluntary organizations, religious groups, NGOs, individual activists and farmers initiated the movement, and they are the ones who are keeping it going. In many ways they have all become "barefoot" engineers and hydrologists who are always in search of some new method or location for capturing more water.

Notable by their absence in this movement are international donor agencies, government engineers, and water experts. Most of them have been skeptical or even hostile bystanders. The same was true for the state government until the recent past; however, from 2000 it has lent its support to the movement. Some critics have described the movement as a mad rush for capturing water runoff, which produces no net improvement in the overall water situation but creates new demands for water in the upstream areas at the cost of downstream users (Shah 2002). The movement's proponents, however, have no time or patience for such criticisms and still proceed with their "mad rush."

Not much is known about the rain catcher movement outside Gujarat. This chapter examines its genesis and primary ideas, its actors, and its achievements. The central question posed is whether its efforts have improved the water situation in Saurashtra – or whether, as claimed by critics, it is all just a "mad rush." Finally, the chapter will examine some of the wider implications that this experience may have for overcoming India's emerging water crisis.

Saurashtra – an overview

The peninsula of Saurashtra, jutting out from the mainland into the Arabian Sea, is the most conspicuous part of Gujarat. It comprises an area of about 60,000 square kilometers and its topography resembles an inverted saucer. This characteristic in particular contributes to the region's water scarcity (discussed below). The region's mean annual rainfall is quite low – about 500 mm, most of which falls over 25 days during the monsoon.[2] Inter-year variability is also very high, about 40 percent. Consequently, rainfall in many years is far less than an "average" year, and in some years it is far greater.

The total quantity of rain water which falls in the region in a normal year is about 30,000 million cubic meters. Much of this precipitation is, however, taken up by vegetation (including agricultural crops) and lost to the atmosphere through evapo-transpiration (referred to as "green water"). Some water does sink into the ground and contributes to the natural recharge of groundwater. The remaining flows through numerous streams and rivers (about 80 river basins) to the sea.

The annual utilizable natural recharge to the ground water in a normal year is estimated to be about 4,500 million cubic meters (GOG 1999) and the surface water potential in streams and rivers – at 60% dependability – is estimated to be about 3,600 million cubic meters (GOG 1996).[3] The drainage pattern is radial, with streams and rivers flowing in all directions away from the centre. Given its topographical characteristics, most of the region's surface water flows quickly through the rivers to the sea; there is not a single perennial river in the region.

Most of the region is underlain by hard rock basalt of the Deccan trap formation, thus it has no deep confined aquifers which could be tapped by digging deep bore-wells. Only shallow, unconfined aquifers (up to the depth of 100 ft.) are found, which are tapped with open dug wells – and these are only replenished with local rainfall. In fact, water levels in the wells rise significantly after rain falls, and then fall rapidly as water from the wells is used for agriculture and domestic requirements. By the onset of summer, most of the wells have little or no water. This seasonal variation in water levels in the wells is characteristic of all areas with shallow, unconfined aquifers.

Scarcity and droughts

Water scarcity and droughts are not new phenomena for the people of Saurashtra. They have always lived with them and thus have treated water as a highly scarce and precious resource. An elderly mother-in-law would always tell her newly-wed daughter-in-law that "It is alright if a jar full of *Ghee* is upturned, but a glass of water should not fall."

Since rainfall is both low and erratic in the region, rain-fed agriculture has always been a very risky proposition, with no security even for *Khareef* (monsoon period) crops. This is the primary agricultural season, when crops are grown using accumulated moisture in the soils from rains. In arid and drought-prone areas with low and erratic rainfall, the security of these crops is often jeopardized if there is no irrigation, and if rain does not fall at the right time. This is why it is said that agricultural production in India depends on the vagaries of rain.

In recent decades, water scarcity and drought meant that less employment was available in agricultural production. A great deal of young people had few options other than to seek work elsewhere, such as Surat and other places in south and central Gujarat where they were employed in the diamond-cutting industry.

Government projects

As in other parts of the state, the Gujarat state government responded to the situation of water scarcity by constructing numerous major, medium and minor dams on almost all the rivers of the region. Nearly 120 major and medium projects have been constructed, with total water storage capacity of about 2,300 million cubic meters in their reservoirs. The main purpose was to provide water through canals to specified command areas for irrigation.

However, since the cities and towns of Gujarat have experienced an increasing need for water, most of the reservoir water in water-scarce areas like Saurashtra and North Gujarat was increasingly allocated to "high priority" urban uses, leaving the farmers high and dry. Farmers might receive water for their winter crops (during the months between December and March) only once or twice in five years.

Somewhat paradoxically, large amounts of water reserved for

summer use in urban areas (and thus denied to the farmers) would often simply evaporate from the reservoirs. Moreover, large parts of the region and the state were not covered by the "command areas" of the irrigation schemes and were left to their own devices for water. All in all, canal irrigation from government constructed reservoirs has not made a significant contribution to the spread of irrigated agriculture in Saurashtra.

Well construction by farmers

Other developments took place parallel to the government's construction of dams on rivers. Left to fend for themselves, the farmers went on a well-digging spree which started in the late 1960s and lasted until the mid-1980s. During this period, the farmers constructed thousands of new open dug wells, and they also deepened old wells.

In the initial years, the wells had ample water — since they tapped into groundwater which had accumulated over the years. However, this water was soon exhausted, especially since water was being used to irrigate both *Khareef* and winter crops. Digging deeper wells was not an option, since the hard rock below does not have any confined aquifers. Still, many people dug deeper bore-wells to capture water trapped in cracks and fissures. But this was a chancy affair; many such wells simply failed. Once the stock of accumulated groundwater was fully utilized, rain water was the only remaining source to replenish groundwater on yearly basis.

In a normal situation, the natural recharge rate of groundwater increases if a majority of groundwater stock has been utilized in the previous year. But given the region's topographical characteristics, with the additional factor that its soils are not highly permeable, this does not happen to any significant degree. Thus, water levels in wells would increase after the rains, but not to the extent required unless there was an unusually high rainfall that year.

This water at most provided the *Khareef* crops with one or two protective waterings when there is no rainfall in the final months of monsoon; the situation would be even worse in low rainfall years, when water is needed most. Finally, a severe drought in 1987 caused acute hardships in rural and urban areas of Saurashtra, Kutch and northern Gujarat. *Khareef* crops failed, thousands of cattle died and there was severe drinking water scarcity, both in villages and urban areas. Water had to be carried to Rajkot (the main city of Saurashtra) in special trains.

The movement begins

After suffering through the drought, there was no way that people would just sit by idly, hoping that the government would take action to change the situation. Everybody started applying their minds towards solving a central question: how could they capture more rain water and put it under the ground for later use? No central organization was created to unify their efforts. Nearly everyone, whether individually or in small groups, addressed the same question and tried out their own solutions. The successful solutions then spread through informal exchanges and site visits. Water then truly became "everybody's business."

Well recharge

Immediately after the 1987 drought, the farmers, individuals and communities, and voluntary organizations started to explore ways that rain water could be captured. The first experiment, tried by some local philanthropic groups and individuals, was to recharge wells. This involved diverting farm run-off and passing it through a filtration bed into the wells through pipes.

Initially, the farmers were very skeptical of the process – but it quickly gained momentum after they saw the real benefits. Philanthropic and religious groups and individual donors provided initial support – in the form of cement and pipes – to willing farmers, and then the farmers started to do it on their own. Within a short time almost all the farm wells in Saurashtra had well-recharge systems in place, and this is still true today. This was done entirely by the private initiative of farmers and local philanthropies. No financial support from government was ever received for this program – nor was it ever sought.

Other experiments

This direct recharge of the wells did help to improve the availability of water, but it was not enough. A great deal of farm run-off was still running into streams and rivers, while groundwater was not being recharged to the extent required.

So again, nearly everyone was engaged in the search for new and better ways to put more water under the ground. All sorts of methods were tested (and are still being tried). These included farm-ponds, percolation tanks, a series of cascading tanks, constructing dedicated recharge wells and check dams. Some methods were successful, and some were not. Some were suitable for specific locations, but not for others. The merits and drawbacks of each method were intensely debated in minute detail – with proponents of any specific method always arguing that theirs was the best!

Check-dams

Eventually, the idea of constructing check-dams caught people's imaginations and was implemented on a large scale. This method relied on the simple idea of building a series of check dams at regular intervals in streams and rivers, approximately 3 to 4 feet in height, in order to obstruct the flow of water in quickly-running streams and rivers. The stored water would then percolate into the groundwater, and eventually a dynamic equilibrium would be established between the two. Specifically, when the groundwater level is below the level of water stored in the streams (behind the check-dams), the stream water will percolate down into the groundwater. Likewise, when the groundwater level meets the stream level, it will flow from the ground into the streams.

However, this idea had to overcome a major hurdle before its implementation was possible on a large scale. Farmers simply had no confidence that they could design and construct such dams: "It requires engineers to decide the location, calculate the flows and then prepare the design," they said. "How can we possibly do these things on our own? What if the dams are washed out away in the first rains?" they asked.

There were a few individuals and groups who had great confidence in the farmers' abilities and were willing to share the risks. They launched a "Cement Support" program which provided the required amount of cement at no cost to those farmers who were willing to construct check-dams.[4] Meanwhile, farmers bore the costs of labor and other materials, and also made decisions about the site location, design and other issues.

Many farmers took up this offer and began to construct check-dams. In fact, their initial experiences greatly boosted their confidence. Some of the dams in the initial phase indeed were washed away – but this was taken in a stride as a part of the learning experience. The farmers quickly mastered the art of low-cost construction of various types of check-dams.

Many other individual donors and NGOs also chipped in and started to support the building of check-dams. Central government funds (allocated under the "Watershed Development Program") were also utilized by the villagers and NGOs for this purpose, giving the activity a tremendous boost. A number of innovative organizations – such as "Vruksh Prem" – insisted that the farmers make a substantial financial contribution (up to 50 percent; normally only a 10 percent contribution from farmers is required). By doing so, they achieved far more than would have been possible under a normal implementation of the program. Most importantly, involving the farmers in this way discouraged the "dependence syndrome" and encouraged a spirit of self-reliance.

In January 2000, the Gujarat state government (under the leadership of Keshubhai Patel) decided to throw its weight behind the initiative. It launched a massive program called the "Sardar Patel Water Conservation" scheme, a version of "Cement Support" initially launched by Vruksh Prem and others. Any group or individual farmers desiring to build check-dams were given support of 60 percent of the estimated cost, while the group or farmer contributes the remaining 40 percent. In total about 24,000 check dams in Saurashtra have been constructed under this program. Given that a great number of check-dams were already constructed before, the total number of check-dams is larger but no one knows what the actual figure would be.

Evaluating the rain catcher movement

The central issue of this chapter is whether the building spree has been worth the effort in terms of improving the overall water situation. The answer, by the movement's protagonists – including farmers, NGOs and local activists – is of course an emphatic "yes." Not only have these efforts augmented the total water storage, but they have also helped to improve utilization of the stored water, they say. A large amount of runoff that previously was lost to the sea or evaporated from the government reservoirs is now captured and stored. Thus, it is contributing towards meeting the growing water requirements of both agriculture and households.

However, various hydrologists and other experts have contested these claims, and have expressed grave concerns about this movement. Their criticisms suggest that "cannibalistic competition" for water run-off will result in down-stream externalities for other users, evaporation losses and a limited potential to recharge groundwater in the underlying basalt.

Down-stream externalities

The primary criticism by many experts is that there has been a chaotic and cannibalistic rush to capture run-off through local water-harvesting structures (WHS), instead of creating additional storage. This criticism suggests that the WHS have only replaced the storage created by the government dams and reservoirs, resulting in new demand for water in the upstream while the costs are inflicted on previous downstream users.

For instance, Dinesh Kumar – a renowned hydrologist with the IWMI – has suggested that "The growing profusion of the small water-harvesting structures captures a bulk of run-off, leaving nothing for the down-stream people who depend upon the government structures for irrigation and domestic water requirements. This not only makes government investments useless but also renders towns and cities of Saurashtra perennially parched" (Shah 2002).[5] The World Bank has also reiterated the point against decentralized water-harvesting programs in general in its latest report on India's water crisis (World Bank 2005).

As mentioned before, the total storage capacity created in large and medium government reservoirs is around 2,200 million cubic meters. The runoff in the streams and rivers of Saurashtra at 60 percent dependability is estimated to be 3,600 million cubic meters, but mean runoff (50 percent dependability) is greater. Hence it is clear that additional storage capacity has been created by the check-dams.

Furthermore, in areas with scarce and erratic rainfall it is always a good idea to create extra storage capacities. If it really were the case that towns and cities are left perennially parched, there would have been huge riots for water in Saurashtra by now. City and town dwellers (and likewise villagers) are not meek folks whose rights can be easily trampled upon. The urban residents continue to enjoy full rights to water from the government reservoirs, and moreover receive water from the Narmada pipeline.

Government reservoirs are indeed filled, in spite of the plethora of upstream WHS. In fact, people in towns and cities also support the check-dam movement. Its beauty is that instead of wasting time and energy in fighting for the limited water that was available, the farmers of Saurashtra silently continued to make more and more water available to themselves in a creative manner. Theirs was a most imaginative response to the crisis, precipitated by an arbitrary government decision to favor towns and cities in its distribution of reservoir water. If there was any "cannibalism" in Saurashtra, that was its real embodiment – and the farmers have responded to it creatively and successfully.

Another "externality" claimed to result from the check-dams is that they have a negative impact on downstream farmers who are located in reach of water from government reservoirs. It is true that they hardly get any water from the government dams. Yet this was true even before the construction of check-dams in the upstream areas, since the government has allocated more and more water to urban users. These farmers, like their upstream counterparts, depend largely on groundwater for irrigating their crops. They too have joined the scramble for rainwater harvesting to augment water availability.

Evaporation losses

Another criticism made by many experts is that given the arid conditions of Saurashtra, large quantities of water stored in shallow ponds behind the check-dams actually evaporate and the people upstream also do not benefit. This argument has little merit and if anything, the reverse is true. Maximum evaporation occurs from the government reservoirs, which do have water in the summer months. According to M. S. Patel, Secretary of Gujarat's water resources department, every year government reservoirs lose 600 million cubic meters (out of a total of 2,200 million cubic meters) of their water to evaporation. This loss is greater than the total annual domestic water requirement of Saurashtra, which is estimated to be 500 million cubic meters (Shah 2002).

In the case of water stored behind check-dams, most of the water percolates down to replenish the groundwater extracted for winter crops – long before the onset of summer. It is the dynamic connection between ground and surface water that the experts seem not to see. In arid areas, surface and ground water must be viewed as a single resource, not separate resources. While the farmers intuitively grasped this idea, the government and the "experts" have generally ignored it.

Limited capacity for ground water recharge

Other researchers have expressed grave doubts about whether these decentralized recharge efforts can produce any beneficial results.

They note the basaltic nature of Saurashtra's hydrogeology, which limits groundwater recharge to 8 to 12 percent of precipitation. They say that a typical hectare of an aquifer in Saurashtra cannot absorb and store more than 400 to 500 cubic meters of water – a small fraction of the amount of water utilized by a single crop in a season. In other words, if Saurashtra were to receive five inches of precipitation, using current recharge methods, aquifers in the region would fill up to the brim. Any remaining water would reappear on the ground as "rejected recharge" (Shah 2002).

If this really were the case, why should we be preoccupied about the upstream farmers capturing all the runoff and rendering government reservoirs useless? Moreover, the state government estimates that natural utilizable groundwater recharge in Saurashtra is 4,500 million cubic meters. (This is equivalent to more than 750 cubic meters per hectare, given that this is utilizable – not gross – recharge.) Here also, the dynamic inter-relationship between the surface and ground water seems to be totally ignored. The fact is that these complex interactions have not been properly studied.

Overall, it seems that the main motivation for concerns about check-dams is their perceived impacts on "planned government projects." At the same time, the experts have totally neglected to study what is actually happening in Saurashtra – to either confirm or deny these perceived impacts, and then to draw lessons from their research. Many have not managed to unshackle themselves from the "planning" mode and are deeply uncomfortable with the farmers' seemingly chaotic and unplanned interventions.

Meanwhile, the farmers of Saurashtra have not waited for such studies and research, and do not even need them. The evidence and the benefits they have observed is sufficient to convince them.

First, they have improved water levels in their wells, and have experienced a consequent increase in agricultural production. Security of the first *Khareef* crop has been assured, not only for years of normal rainfall, but also for years of scanty rainfall. Even one or two good rain showers are sufficient to capture enough water to protect the *Khareef* crop. In normal years, winter crops are also assured.

Second, the growing agricultural sector in the area – which previously experienced outward migration – now provides employment to migrants from Panchmahal and other tribal areas. Agricultural wages have risen; the daily wage is as high as Rs.100 (approximately USD $2) during the peak season, a relatively high wage for agricultural workers.

Third, the previous crisis over drinking and domestic water has simply disappeared.[6] Moreover, in good monsoon years there is even surplus water at the end of year, which can be utilized the following year. It is possible that these benefits have not been experienced in all of the region's localities – but the overall effects of the rain catcher movement can certainly be observed while passing through the area.[7]

Wider implications

It is clear that Saurashtra is not just a quirky exception; there is no reason why similar local interventions to address water scarcity in other areas of India should not work with equal success. Experiences from similar areas in Rajasthan and Madhya Pradesh indicate that such local interventions are indeed helping to resolve water issues.

However, this intervention of local, decentralized rain water harvesting would be unlikely to work in all areas. For instance, it will not work in arid areas – such as Mehsana and Banaskantha in the northern Gujarat plains – where tube-wells are used to mine groundwater from deep confined aquifers. In fact, the farmers of northern Gujarat are so convinced that local water-harvesting schemes will not work that they have not even attempted to implement them.[8] Water scarcity in these areas is indeed difficult to tackle and would require some difficult choices – but that is beyond the scope of this chapter.

Fortunately, the conditions are contained to specific locations and largely do not prevail in the rest of the country. In all other water-scarce, arid areas, decentralized interventions to harvest rain water and recharge groundwater – appropriately adapted to local situations – can make a significant difference.

Apart from this, the Saurashtra experience also raises some critical questions for policymakers, especially regarding some of the fashionable characterizations of "market-oriented" policy solutions which are often advocated by organizations such as the World Bank. A few of these are discussed below in the light of the Saurashtra experience.

Solution 1: Most of water is used by agriculture; minor reductions in this use can release a substantial amount of water which can and should be used to meet the growing urban needs.

This is a major misconception, at least, in Saurashtra, in northern Gujarat and in most of India's water scarce areas, especially where water is not imported from other regions which have a surplus. Here, most of the surface water stored in government-constructed reservoirs is already allocated to the cities and towns, while farmers get less, and in an unreliable manner.[9] During years of scarcity, water supply for irrigation is drastically cut, without any compensation to the farmers. This has caused severe hardships. Further reductions would need to come from reducing use of groundwater, causing further hardships and making agriculture itself an unviable proposition.

This solution also does not take into account seasonal factors. It is true that agriculture requires substantial quantities of water – but the bulk of this use in India and south Asian countries is during the monsoon and in winter, when towns and cities usually have surplus water. Agriculture in water-scarce regions uses very little (if any) water during summer months when cities face scarcity (Frederiksen 1996). In Saurashtra, we have seen that in order to reserve a large proportion of water for urban areas for summer use, governments have denied it to agricultural users in the winter. Yet a large amount of water evaporates from the reservoirs, without providing benefits to anyone.

So how can we meet the growing needs of the cities? This is a difficult question and there are no easy solutions, especially in areas where water is already scarce. More water will need to be made available. Decentralized rainwater harvesting is definitely an important option, as we have seen in Saurashtra. The option of long distance inter-basins transfers also has to be explored. In Saurashtra too, despite the water harvesting structures, water availability for cities would become critical in scarce years if the region only relied upon its own water. The availability of Narmada reservoir water through pipelines has provided long-term security for the growing requirements of the cities in Saurashtra. The main issue is that with growing population and increasing needs, we need to increase the amount of water available for cities as well as agriculture.

Solution 2: A system of water use rights for surface and ground water should be established. This would encourage markets for water rights and would facilitate voluntary transfers of water from agriculture to cities thus ensuring long-term sustainable use of water.

A full discussion of this currently "in vogue" solution is beyond the scope of this chapter. Instead, a few critical issues are raised here in light of Saurashtra experience.

Many proponents of this solution may not have any idea of the immense complexities and difficulties involved in its implementation. This solution proposes to create a system where annual use rights for each and every individual user of surface or ground water are determined, allocated, recorded, monitored and enforced. These rights are to be determined with criteria based on present uses (for instance, average use in last five years). If the aggregate present use is more than that can be sustained on a long-term basis, then necessary deductions will be made.

This idea would present several complexities that relate to inter-year variability of rainfall. Adjustments would also have to be made in the amount of water that can be used, based on rainfall in particular year. Also, there would be issues relating to rights to return flows. The total water diverted by each user is not fully used in a consumptive manner; some is used and the rest flows back into the streams or percolates down to the groundwater, and is used by others downstream. These rights to return flows would also have to be determined and recorded. Finally, such a system would require methods to monitor actual use (as to whether use rights have been respected or violated). It would also require some form of punishment for any offenders.

Although couched in terms of "property rights" and "markets", this solution actually advocates government regulation and quota-fixing for all individual users of water. No state government in India is in a position to implement such an ambitious scheme even if it wanted to do so, but that is not the main point. The main issue is whether or not the creation of such a system would help in solving the looming crisis of water scarcity.

Here, the experience from Saurashtra suggests that far from solving the water crisis and associated conflicts, this solution would actually indefinitely delay the search for real solutions. It would result in a fruitless waste of time, energy and money – and a lot of bickering. Moreover, it would not result in a net gain in available water.

The situation in Saurashtra would be very different today if the farmers – instead of engaging in decentralized experimentation and adopting methods which made more and more water available – had clamored for their own rightful share and fair distribution of the total quantity of water which was previously available. Their experience shows that what is needed is to increase the total size of the pie, making more water available for everyone, rather than creating conflict about the distribution of the present pie!

Those who advocate such solutions often forget that property rights for many resources evolve spontaneously in actual practice, if governments do not interfere. Such institutional arrangements – the set of rules by which individuals and communities allocate their resources – might then be formally recognized by the law, but not vice-versa. The fact that a form of individual, permanent and transferable property rights for water use have not spontaneously evolved in large parts of the world indicates that there is something in the nature of water which prevents evolution of such rights. [10] The British common law jurist William Blackstone remarked that "Water is a moving wandering thing, and must of necessity continue common by the law of nature; so that I can only have a temporary, transient, usufructuary property therein."

Following from the British common law doctrine, India legally has the doctrine of riparian rights, where water which flows through a river basin is the common property of all residents of the basin, and each such resident has a right to reasonable and beneficial use of water flowing through their area without prejudicially affecting similar rights of other residents.

Similarly in the case of groundwater, it is the common property of all the overlying landowners of the groundwater basin. Each land owner has a right to extract and use water that lies below his land, without prejudicing the rights of other landowners. This is the legal position, and it is how people have traditionally used water, not only from streams, rivers and wells, but also from tanks and canals.

In actual practice, however, state governments have acquired monopoly rights over surface waters during the past fifty years. This has been achieved not by any change in law, but simply by virtue of the fact that most of the dams, reservoirs and canals used to divert river waters were constructed and are managed by the governments. Governments rarely involve local communities in planning, execution and management of such projects and their general performance has been very poor. Not only have they often seriously undermined the riparian rights of the basin users, but they often arbitrarily changed allocations of the stored water.

The "rain catchers" of Saurashtra broke the government monopoly on surface waters by reestablishing their riparian rights. They have done so in the same way in which government acquired its monopoly rights – by creating their own water storage structures. In the process, they have increased the overall storage capacity and thus have improved water availability for all users. All these farmers are very much aware that they are re-defining rights and not just building check-dams. One of their primary slogans (which is also backed by broad popular support) is that "Water belongs to those on whose lands it falls and they have the right of first use of this water."

Rather than a "top-down" system which is created and imposed from above and has no social sanction, this decentralized, evolutionary process by which riparian rights are re-established needs to be encouraged and supported.

In the case of groundwater, the riparian rights of overlying landowners have in actual practice remained in force, simply because private efforts by farmers (not governments) have driven groundwater development. Each overlying landowner has an absolute right to extract and use the water which lies below his land. This right is recognized and respected by all. Of course there were problems of "free-riding" in the initial stages of development. This was especially the case when only relatively wealthier farmers could afford to dig wells or tube wells which enabled them to extract not only their own water but also that under neighboring lands.

However, this problem was solved by the neighboring farmers in a simple and straightforward manner. Whether as individuals or collaboratively in the form of partnership companies or cooperatives (which helped to resolve finance issues), they dug their own wells. This automatically reduced the "extra" water that could be extracted by the previous well owners. Water markets have also developed, enabling those without wells to use water underneath their land by purchasing it from those who do own wells.[11]

Perhaps the most striking feature of the Indian situation is the lack of undue worry on the part of farmers about the problem of "free-riding." They have never felt the need to monitor and restrict the amount of water extracted by each well owner, either by forming their own organizations or by seeking government intervention. By the time that groundwater development reaches maturity, as has happened in most water scarce areas, the amount of water extracted by each owner is not much different from what others are extracting.

Governments have, of course, tried to exercise control over this "chaotic" development by passing various restrictive laws. But mostly these remain laws on paper and have never been seriously implemented. It should be noted as well that digging wells and extracting groundwater has never been considered an "illegal" activity as such. Privately-driven development of groundwater now is responsible for 70 percent of the net irrigated area of the country, and also provides water for domestic uses to cities and other nearby villages in times of acute scarcity.

Groundwater has performed far better than that of the government-driven canal irrigation sector. Moreover, unlike canal irrigation, the use of groundwater has become highly efficient. The use of water-saving devices, such as drips and sprinklers, has encouraged a shift towards high value but less water-intensive orchard crops; this shift is picking up momentum with farmers. So with groundwater too, there is hardly any need to establish micro-managed quotas and enforcement systems. Far from reducing groundwater use, this would only create diversions and bickering.

Strangely enough, proponents of a so-called "market solution" for water seem to be uncomfortable with decentralized water harvesting by farmers, in which the farmers have reasserted their riparian rights. Instead, those proponents (who sometimes refer to their solution as the establishment of "individual, permanent and transferable property rights" for water) seem to be deeply suspicious of unregulated, uncontrolled markets and decentralized institutional arrangements. Thus, they actually seem to advocate a tremendous increase in the role and powers of government over water allocation and use. True proponents of markets should be extremely wary of these "neo-marketeers."

Notes

1. This movement for decentralized rain water harvesting and ground water recharge is occurring in both Saurashtra and Kutch, two water scarce regions of Gujarat. The situation, interventions and impacts are, however, different in these two regions. This chapter only discusses the Saurashtra experience.
2. Monsoon is a characteristic phenomenon of India and other south Asian countries. It is a period starting from the beginning of June to end of September during which most of the rain from south-western winds falls. This is also the main growing season for most plants and crops.
3. The estimates of groundwater recharge and surface water potential given here are of utilizable quantities. The figures for gross groundwater recharge and surface water flows would be higher than that indicated by these figures. It may also be mentioned that no systematic studies on groundwater recharge have been done. The state government's estimates are more often than not the result of "back-of-the-envelope" calculations based on total area, rainfall, soil types, etc. The recharge which occurs in areas with saline groundwater is then deducted.
4. Two individuals played a pioneering role in the movement by conceiving and launching the "Cement Support" program with their own money. One was Premjibhai Patel, founder of "Vruksh Prem", a local voluntary group in Upleta, and the other was O. R. Patel, an industrialist of Morbi who manufactured famous wall clocks (Ajanta Clocks). At the time, few others had confidence that farmers could successfully build check-dams. Premjibhai supported construction of about 500 such check-dams from the money he raised from individual donors (mainly from his own entrepreneurial/industrialist son). Both he and O. R. Patel have given away cement worth Rs. 50 lakhs each [in present 2006 terms this is worth over USD $113,000] for water conservation works in general. In addition, Premjibhai has also helped to construct nearly 1500 check-dams in about 40 villages under the government's watershed development program. My understanding of the genesis and evolution of the 'rain catcher' movement is largely based on my long discussions with Premjibhai and others who have played active role in this movement.
5. Also based on personal discussion with Dinesh Kumar.

6. I am referring to the hinterland areas of Saurashtra, and not coastal areas which still suffer from the problem of sea-water intrusion in the groundwater. The hinterland villages are not only meeting their own requirements, but are also providing water through tankers to meet the domestic requirements of these coastal villages. The area has a large number of water traders who buy water from farmers in the hinterland areas, and then sell it to coastal villages through tankers. This has become the main source of domestic water supply in many coastal villages which have saline groundwater.
7. I am mainly relying on my observations during various field visits in 40 villages where Vruksh Prem has been working. These are in the upstream area. However, similar evidence from across the Saurashtra suggests similar changes. Tushaar Shah has presented and discussed some of this evidence in his paper (Shah 2002).
8. Although the state government, as usual, is promoting check-dams in all water scarce regions of the state, irrespective of the local conditions.
9. In humid water-rich areas, for example in south and central Gujarat, irrigation receives a major portion of the stored reservoir water. However, towns and cities in these areas also get their share of water, which is adequate to meet their requirements and thus they do not face any water scarcity.
10. Perhaps in actual practice, the water scarce western region of the United States is the only region in the world where such rights evolved under the doctrine of "prior appropriation" and were then accepted in law during late 18th and early 19th centuries. Discussion of how this happened and whether they have led to voluntary transfer of water rights from agriculture to cities as expected or not is interesting but beyond the scope of this chapter. See Morriss (2002) for more discussion of this issue.
11. Large numbers of well-owners have ensured competition and thus have not allowed "monopoly pricing" to take hold. Shah (2005) discusses this and many related issues in terms of India's water management.

References

Frederiksen, Harald D (1996). "Water Crisis in Developing World: Misconceptions About Solutions." *Journal of Water Resources Planning and Management*, Vol. 122, No. 2 (March/April), pp.79–87.

Government of Gujarat (GOG)(1996). "Water Resource Planning for the State of Gujarat – Phase III." Vol. II, Main Report.

— (1999). "Report of the Committee on Estimation of Ground Water Resource and Irrigation Potential in Gujarat State: GWRE – 1997." Narmada and Water Resources Department, Gandhinagar.

Morriss, Andrew P. (2002). "Lessons from the Development of Western Water Law for Emerging Water Markets." *Oregon Law Review* vol.80, pp.861–46.

Shah, Tushaar (2002). "Decentralized Water Harvesting and Groundwater Recharge: Can These Save Saurashtra and Kutch from Desiccation?" Paper presented at the annual partners' meeting of the IWMI-Tata Water Policy Research Programme, Anand, February 19–20.

Shah, Tushaar (2005). "The New Institutional Economics of India's Water Policy." Paper presented at workshop, "Africa's Water Laws: Plural Legislative Frameworks for Rural Management in Africa", 26–28 January, Johannesburg, South Africa. Online: http://www.nri.org/waterlaw/AWLworkshop/SHAH-T.pdf (visited 14 February 2006).

World Bank (2005). *India's Water Economy: Facing a Turbulent Future*. (Also referred to by World Bank as *India's Water Economy: Bracing for a Turbulent Future*). Washington, DC.

6 Water governance in China: The failure of a top-down approach

Wang Xinbo

As China's economy grows and expands, an ecological crisis also looms. The most serious environmental issue is that of water. In northern China there is a popular saying: "all rivers dried up, all water polluted". Water scarcity mainly affects the northern region, but water all over the country is polluted to some extent. In fact, the water scarcity in Northern China was the impetus behind the South-to-North Water Diversion Project, the largest and longest project of its kind in the world.[1]

What has caused these problems? What can the Chinese government do about them? This chapter outlines three aspects of China's water governance, in terms of quantity, quality and service. It analyzes the shortcomings of the current system, which allocates water resources and manages water quality through a "top-down" approach that relies on centralized political mechanisms, which is identified as the fundamental cause of the country's water problems.

The chapter also analyses difficulties encountered by the Chinese government in market reforms of the water service industry. Problems which have occurred in this area – particularly with public-private joint ventures – are also the result of the "top-down" approach.

Water is central to China's economy. Whether the country can sustain its high growth largely depends on whether it can transform

148 The Water Revolution

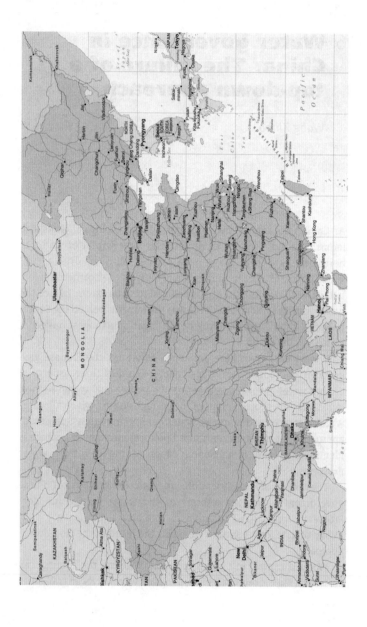

the current system of water management into one which relies upon decentralized, "bottom-up" decision-making. This means decision-making rooted in individual property rights, and other mechanisms, such as water user associations, that enable individual and community decision-making.

Water quantity governance

Geographical background

China uses only one-fourth of the average per capita water of the whole world. It is considered to be one of 13 water scarce countries by the United Nations. The basin area of the water systems in northern China (north of the Yangtze River) covers 63.5 percent of the total land area, yet its aggregate water accounts for only 19 percent of the total of the country. Inland river basins cover 35.3 percent of the area in northwest China, yet the region has only 4.6 percent of the country's aggregate water.

The problems of long-lasting drought and water scarcity are especially serious in northern China's three-H rivers region (the Huang, Huaihe and Haihe Rivers) – an area characterized by many contradictions in terms of supply and demand. Here, water per capita amounts to only one-fifth of the average level of the whole country. Yet water resources in the three-H rivers region have been developed and utilized at rates as high as 80 percent, 65 percent and 98 percent respectively. This rate of utilization far exceeds sustainable levels of approximately 30–40 percent for those rivers.

At present, the groundwater in Northern China has been over-exploited, and the groundwater level has fallen rapidly. For example, the groundwater level of Gongzhufen area in Beijing has been descended to the bedrock and has been exhausted. At present consumption rates, the groundwater of Northern China will be exhausted in the next decade (Wang Shucheng 2003).

Paradoxically, although China's water is extremely scarce, water use efficiency across the country in general is extremely low. For instance, the amount of water required to create additional industrial output value of 10,000 yuan (US $1,242)[2] is 5 to 10 times that of the developed countries (whose water utilization rates are only 40–45 percent).[3] This demonstrates a great potential and scope for enhancing water efficiency in China. Currently, however, these gains in efficiency are elusive, since there is a lack of arrangements that would enable decentralized decision-making.

Water resources management

The existing systems to manage water resources comply with the Water Law of 1988 and the newly modified Water Law of 2002. Water management in China can be divided into two periods. Before 2002, water abstraction was managed through a license system. Since 2002, when the new Water Law was enacted, the country has been making a transition to a rights-based system.

Before 2002 – the license system

The 1988 "Water Law of the People's Republic of China" clearly defined the ownership of the water resources and established a license system, as well as a system of user fees, for water abstraction. The government subsequently issued various regulations relating to the procedures, supervision and management of the license system.[4] At the same time, regional regulations[5] – relating to the collection of water use fees – were formulated and executed in some areas.

The license system is based on the fundamental idea – stated in Article 3 of the 1988 Water Law – that water resources are owned by the state. Thus, the state allocates and manages water with a hierarchical administrative system.[6] The Water Law established a top-down allocation mechanism, through which the State Council enforces the State's ownership. The Ministry of Water Resources (MWR), an administrative department, was established to perform this task. The MWR allocates water management to lower-level government agencies. It establishes branch organizations in the river basin or region, and develops a hierarchical administrative management system.

Each administrative region carries out a "Total Quantity Control" for its water resources, according to a plan of the river basin. Both surface and ground water resources are allocated to the provinces, cities and autonomous regions within the river basin. The provinces, counties and cities have examination and approval power over their allocated quotas. The MWR is responsible for examination and approval of any license above and beyond the allocated quota for each large river basin and large reservoir.[7] Finally, a water use fee is collected for both surface water and groundwater. Revenue from the fee is divided by financial departments at different government levels and then allocated to administrative water departments.[8]

How does the "top-down" approach work in practice? During the period in which water was managed with the license system, government at all levels did not restrict the total water consumption. It unintentionally gave a green light to a "pump race", which caused a series of problems such as dried-rivers and ground subsidence.

The top-down approach is based on several false assumptions. First, it assumes that each level of government has accurately estimated the national and local surface water and groundwater resources, and can respond in a timely fashion to any changes of water resources or other factors. Moreover, it implies that government has a clear supervision over each user's actual water use and, especially, any excessive water abstraction.

Second, it assumes that the state will be able to enforce restrictions on the total water consumption of the subordinate government and water users.

Third, it assumes that the administrative authorities can, and will, faithfully safeguard the public interest by turning a deaf ear to the rent-seeking attempt of water users.

In reality, there are several reasons why these assumptions do not pertain to the existing water management system. First, the data utilized by the state – such as basic data pertaining to local water resources, infrastructure, and the amount of water consumption – is outdated. Data provided by the lower-level administrative departments is generally out-of-date and inaccurate.

Secondly, government agencies and departments always satisfy their own water requirements. The government presently lacks an overall assessment of water resources, and thus cannot enforce restrictions on the total amount of water abstraction. When such an assessment exists, the government tends to overestimate the total amount of water resources, and underestimates actual water consumption. At the same time, the upper levels of government have great difficulty enforcing restrictions on their subordinates. So, the permitted amount of water always exceeds the available amount of water resources. While this does serve to increase the power wielded by public officials, these deficiencies prevent successful management of water resources.

Third, it is extremely costly to monitor water abstraction, because of the highly dispersed water resources and cross-penetration between the main stream of the river and its branches, as well as between surface water and groundwater.

Fourth, water administrative authorities are easily tempted and captured by interest groups. Local governments are hostage to local economic interests, including state-owned enterprises, the private sector and foreign investors. Local governments become tools for rent-seeking, by helping politically powerful water users to use more water while paying less in fees. Moreover, governments arbitrarily use their power to reduce the fees or even exempt them from payment. Indeed, in some areas such exemptions are used as a preferential policy to solicit investment.

The coordination costs of the license regime have also been extremely high. For instance, the system of water use fees has been very costly to implement and has many shortcomings, according to a survey by the Department of Water Resources of the MWR. Some users who owe fees instead issue "IOUs" and delay their payments as long as possible.

Meanwhile, some enterprises (in particular, those with poor returns) are allowed to default on their payments. A lack of proper metering equipment makes the situation worse. It is nearly impossible to measure actual quantities of water abstracted, which is the basis of the relevant fee. Some local water departments intentionally neglect to collect the fee for self-owned wells, which has contributed to excessive exploitation of groundwater. Because it encourages local officials to engage in discretionary use of their power, the license system has also encouraged corruption.

Fifth, all levels of Chinese government not only oversee public policy, but also invest in many state-owned enterprises. Their primary goal is to develop the local economy, and to maximize GDP and their own revenue. Even though local government is the manager of water resources, it is also a consumer. This is a conflict of interest, and leads government to actively seek to maximize local water consumption.

Sixth, if excessive abstraction of water is very popular, no user's rights can be guaranteed.

These observations show why the license system defaults towards over-exploitation. The public at large has few opportunities to influence government's decisions over local water management. This means that a license to abstract water simply becomes meaningless words on a piece of paper.

As a result, the license system has exacerbated problems of water scarcity especially in regions such as northern China, where dried-up rivers, excessive exploitation of groundwater, and ground subsidence are commonplace. The "pump race" encouraged by the "top-down" approach has led to a large number of disputes over water affairs, even violent conflicts. [9]

Introduction of water rights

The logic of water rights is that a decentralized, "bottom-up" approach creates superior outcomes to a top-down approach.[10] To function properly, water rights require the adoption of decentralized, bottom-up institutional arrangements. Water rights are a form of property, and it is the major function of government to protect property rights.

In addition to being protected as property, decentralized water management might involve water user associations and the public more generally. Whatever other features it has, it requires the development of a legal framework that would enable transactions to occur between users. In such a system, the higher level of government would play only a coordinating role, and problems would generally be addressed through self-governance. Problems that would ultimately be handed over to higher levels government are those of an external nature.

In China, "water rights" refer to rights of things based on state ownership of water resources. Such rights are usufructs, developed under legal restrictions and with certain limitations. Currently, there is no clear legal definition of usufruct in China, but the Property (*Jus in Re*) Law currently being drafted has listed water rights as usufructs.

China's present water management system is undergoing a slow transition to water rights. A breakthrough has occurred since 2000 in terms of development of rights, both theoretically and practically. In 2002, the Water Law was revised and enacted. The newly-revised law reiterates the state ownership of water resources, defines the water use rights of the collective agricultural economic organizations, adds some control systems such as water amount allocation system and the combination of the Total Quantity Control and quota management, and sets up water abstraction rights.

The government is now devoting effort to strengthening the Total Quantity Control Program, so as to protect water rights and boost the stability and transferability of water rights. Over recent years, China has also issued a series of laws and regulations which complement different aspects of the water rights system. Now the revision of the "Implementation Measures for the Water Abstraction License System" is being carried out to further meet the requirements of water rights management, and the market mechanism will play a more important role in allocating water resources. Some river basins suffering from water shortages have formulated their own water allocation plans according to related laws and regulation.

There have been a few voluntary or government-guided water

transactions and water rights transfers between different regions or users. Some examples of transactions include:

(1) A water use transaction between Dongyang City and Yiwu City. Permanent use rights of 50 million m^3 of irrigation water were transferred by Dongyang City to Yiwu City (for urban use). This was the first practice of the transfer of a water use right in China.[11]
(2) In 2001, 30 million cubic meters was transferred from Zhanghe River in Shanxi Province to Henan Province. In 2002, the two provinces transferred 30 million m^3 between each other.
(3) In 2004, under the guidance of the MWR and the Water Resources Commission of the Yellow River, Ningxia and Inner Mongolia carried out a combination of investment in water conservancy and transfer of water rights. Now eight large industrial projects have signed up to water rights transfers with the irrigation areas. This has promoted better use of water resources between different industries.[12]

Overall, the awareness of water rights has increased, and this plays an important role in monitoring and curtailing inappropriate government actions. The introduction of water rights is a significant step forward – yet it is severely restrained by the existing power hierarchy which resulted from the license system. The Chinese government's approach to water rights still carries a flavor of centrally-planned allocation of resources.

Thus, the state believes it must strengthen top-down planning of water resources in order to strengthen water rights. Yet the logic of decentralized water rights is incompatible with the top-down, hierarchical administration perpetuated by the license system. Until now, except for a few successful pilot programs, the top-down allocation system has not been transformed. This creates uncertainty as to whether or not the system will actually evolve into true water rights.

Wang Yahua (2005, 316–317) has outlined how China might

make the transition from the water license system to a system of water rights:

> *The low-quality water rights held by communities and individuals are often legally defined as "the administrative license" (a privilege). They are political rights, which are allocated and reallocated mainly by administrative means. With various restrictions and interventions, the government has undermined the utility of such rights.*
>
> *To improve the quality of the right, water rights must become usufructs (licensed property right). If the quality of the right can be further improved by the strong protection of the government, it can even become an absolute property right similar to ordinary assets.*
>
> *With the decrease of government intervention and restriction, the water rights will become an economic resource, and increasingly will be allocated through the market.*

Water quality governance

Background

The State Environmental Protection Agency (SEPA) is responsible for administering water quality in China. The approach to controlling water pollution is quite similar to the top-down approach to water resources management.[13] It relies on five hierarchical levels of administration: (1) total amount of wastewater emissions; (2) wastewater emission permit; (3) wastewater emission fees; (4) wastewater emission monitoring; (5) assessment of environmental quality.

In spite of attempts by the state to control pollution, over 50 percent of the river sections of the seven large water basins areas in 2004 have been polluted to different degrees and the water quality of 28 percent of the river sections has made them unusable. Water pollution accidents have occurred frequently, causing serious economic loss. Statistics show that 3,988 accidents occurred from

2001 to 2004 across the country, with almost 1000 on average per year (NPC Information Center 2005).

A lack of access to safe drinking water is also a serious problem across the country. According to the investigation and analysis of the MWR, 250 out of 1000 surface water sources are disqualified for use as drinking water. In rural areas especially, 300 million people still lack access to safe drinking water.

Some enterprises have not abided by the law, and their problems of excess pollutant discharge are quite serious. Some enterprises have built wastewater treatment facilities, but the facilities do not operate continuously. It is well-known (but seldom acknowledged) that these facilities are often used just for inspections, and are turned off when the inspectors leave.

The following two examples provide a clearer picture of China's attempts to manage water quality.

Example 1: Pollution control of the Huaihe River

The Huaihe River is China's third longest river, and is home to one-sixth of the country's population. The river and its many tributaries provide drinking water for many of the towns, cities and villages located in the river basin.

Many industrial enterprises are also located in the river basin, including pharmaceutical, pulp and paper, chemical, beverage, textile and food industries. Industrial waste water, combined with agricultural run-off, has contributed to high pollution levels in the river.

Moreover, the cities along the river have extremely poor sewage and wastewater treatment. The valley has experienced growth in population, which has led to more generation of sewage and wastewater. Zhang Zhenhai, the chief of the Environmental Monitor Station of Zhoukou city in the province of Henan, estimates that more than 85 percent of the sewage produced by small and medium-sized enterprises in Zhoukou does not flow into the river, but instead evaporates or infiltrates the ground instead.

A significant proportion of enterprises dispose of sewage in this way. The sewage infiltrates the groundwater. Because surface water and ground water are connected by the same circulation system, the groundwater becomes polluted, and then the sewage circulates into the river. The damage in terms of human and environmental health can be long-lasting.

To address the pollution in the Huaihe River and its tributaries, the state promulgated the "Interim Regulations Concerning the Prevention and Control of Water Pollution in the Huaihe River Valley Water" (the only valley environmental protection code of China), and both the Ninth and Tenth Five-Year Plans for Prevention and Control of Pollution in the Huaihe River Valley. Four provinces in this valley – Henan, Anhui, Jiangsu, and Shandong – also passed relevant laws, regulations and programs to control pollution of the river.

Over the course of a decade – between 1994 and 2004 – over 60 billion yuan (US $7.2 billion) was spent by state and local governments and the private sector in clean-up efforts. Following this decade of 'pollution control', the Environment and Resources Committee of the National People's Congress organized the "Chinese Environmental Protection Survey" in August 2004. Journalists from the Xinhua News Agency were sent to conduct a thorough investigation about water pollution, especially as to whether the government's pollution control efforts had succeeded along the river.

About one month after the journalists' investigation, a serious pollution rebound was found in the river. The main water quality indices reached or even exceeded the historical record levels. Pollution extended from the surface underground, directly affecting the lives of 130 million inhabitants. Measurements in 2004 indicated that about 60 percent of the water in the valley is low quality (Class V or below), and cannot even be used even for low-grade industrial purposes (State Environmental Protection Administration 2004).

In July 2004, Pan Yue, the deputy director general and news spokesman of the State Environmental Protection Administration delivered a speech on the extraordinarily serious pollution of Huaihe River. He acknowledged that the Huaihe River watercourse had basically lost its ability to self-purify, and that a decade of pollution control efforts had failed to improve the river's pollution levels. The SEPA also closed 52 factories in June 2004.

The decade-long pollution control effort in the Huaihe River has triggered an ongoing debate, and event violent arguments, among the news media (*China Economic Times* 2004).[14] Officials from SEPA like to emphasize their outstanding achievements, and are inclined to be optimistic about the environmental quality, but other departments are apt to criticize it.

The latest debate is about the information promulgated by the environmental department, on one hand, and the water resources organization, on the other hand, about the Huaihe River water quality and related pollution control.

In 2004, the SEPA measured 700,000 tons of COD discharge (the total chemical oxygen demand, which is the key water pollutant index). This indicated a reduction of 50 percent in COD levels since pollution control efforts began in 1994, so the SEPA concluded that pollution was "decreasing year by year."

In contrast, the Huaihe River Water Resources Organization (which is subordinate to the Water Resources Ministry) determined that in 2003, the annual COD discharge amounted to 1,230,000 tons. It concluded that pollutant levels had not changed at all, despite a decade-long effort by the state to control pollution.

A public outcry ensued in response to this information. Further investigation by relevant specialists revealed that SEPA and the water resources organization used different methods to measure pollution. While the former only surveys water quality in mainstream water courses, the latter measures water quality in the factory drains. The former calculates based on a 225-day for a whole year while the latter calculates 300 days. These differences showed that SEPA underestimated COD levels by about 50 percent.

Example 2: Pollution of Songari River, November 2005

In November 2005, the explosion of a bi-benzene factory of the Jilin branch of the China National Petroleum Corporation (CNPC) formed a pollution agglomerate as long as 80 kilometers on the Songari River. What matters for the purposes of this chapter is not why the bi-benzene factory exploded, but how the authorities treated and dealt with this significant event.

The pollution on the river stretched for more than 1000 kilometers, an area covering hundreds of cities, towns and villages which rely on the river for water. The area is both industrial and agricultural, so the incident also affected these users. Yet from the beginning, the public was not informed about the event. In fact, the leaders of the CNPC branch "brazenly hid the truth from the media and even formally prohibited its staff from leaking relevant information" (Hu 2005).

Timeline of Harbin incident, November 2005[15]

13 November – Explosion occurs at CNPC plant.

18 November – Five days after the explosion occurred, the provincial government of Heilongjiang was informed of the event. Emergency work at the earlier stage was led by Jilin provincial government (located on the contaminated upper reaches), and the State Environmental Protection Administration was in charge of evaluation, suggestions, and also coordination between Jilin and Heilongjiang.

21 November – Along the lower reaches of Songari River, Harbin (a city of 10 million residents) was the city which was most seriously affected by this contamination, since nearly all of the city's drinking water comes from the river. In order to ensure that water was safe for human use, the municipal government decided to cut off the water supply for four days. Yet its stated reason – that the "water supply pipe network was under repair" – was farfetched and unreasonable. The masses were aroused to harbor suspicions and an atmosphere of panic resulted.[16]

22 November – In the evening, Harbin's municipal government issued an announcement about the fact that the Songari water might be polluted after the explosion. People were set slightly more at ease, enabling an orderly situation to ensue.

On the same day, the neighboring country Russia was informed of the event and the pollution. The province of Heilongjiang borders Russia, and Songari River flows into Heilongjiang River along the Sino-Russia border. With good reason, Russia has been quite sensitive about the Songari water quality. Information about Songari pollution first went out from Harbin. The Russian media issued numerous comments about it, and then the relevant department in Russia issued a formal notice.

24 November – In the morning, Xie Zhenhua – the director general of the State Environmental Protection Administration – had an interview with the Russia ambassador to China, and communicated with him at length. SEPA officials emphasized on the news release of that afternoon that there were still about 14 days for the pollution agglomerate to reach Heilongjiang River: "Judging from the present situation, it is weakening, and its influence will become smaller and smaller," therefore this notice "was not late."

24 November – After the news release, the provincial government of Jilin started immediately to implement the emergency plan, and the State Environmental Protection Administration also went into action.

28 November – Drinking water was restored to Harbin, 15 days after the explosion occurred.

9 January 2006 – The State Council issued a "General Contingency Plan for Public Emergencies in China". It stipulates that the news of a public emergency shall be released to the public in the first instance. This means abolishing certain related regulations – such as the "Notice on Strictly Enforcing Disciplines of Reporting the

Environmental Pollution Accidents" – which requires such events to be kept secret.

Eventually, the Director General of the State Environmental Protection Administration took the blame for handling this event unsuccessfully, and resigned. There is still much clean-up which needs to take place.

Evaluation of water quality governance

The issues that affect China's management of water quality are the same as those that create its inability to manage water quantity. These boil down to the reliance on top-down administration, which in turn depends on the state's capability to collect the right information. The systems rely on a sense of responsibility of superior government officials. Both systems are fixated on micromanagement and quotas.

If the operating costs of this system were very low, it might be logical and acceptable. Overall, China's laws and regulations related to water pollution control are nearly perfect on paper, but in practice, their enforcement costs are huge. There is a high cost to obey and enforce the law, but a low cost to breach the law. Judging from the example of Huaihe River, the overall water quality of this valley has not been improved despite expenditure of billions of Yuan; this is verified by serious pollution and dry rivers. Thus, top-down administration of water quantity and quality is both costly and inefficient.

Moreover, the laws pertaining to water quantity and quality are not enforced across-the-board. This is especially true at a local level: 'Local protectionism' leads state officials to use political discretion which favors local enterprises. This might entail reducing their water use and sewage fees, which in turn contributes heavily to low collection rates and cost-recovery. It could mean shielding enterprises which engage in illegal sewage discharge, or enabling them to evade supervision and monitoring. Moreover, since local governments pursue local economic benefits and political achievements, they are inclined to approve new industrial projects without proper consideration of their overall effects on humans and the environment.

The system easily lends itself to embezzlement, bribery and corruption of public officials. When a law is easily broken, or an unlawful practice is not easily detected, or if violations are, the law means nothing – it is only a law on paper.

Mao Rubai, the director of the NPC Environmental Resources Committee gets to the truth with a single pertinent remark:

> *The reason why environmental pollution in some places cannot be solved for a long period of time is ultimately traced to the government. The government does not administer by the law, nor does it enforce it strictly, nor does it manage state affairs in accordance with the law. Consequently the environmental protection law is nothing but a scrap of paper.*

Governance of water services

Regulation framework

China's governance of water services retains many features of the planned economy. No single organization maintains uniform supervision and control over water services. At present, water source projects are the responsibility of local water resources departments. Distribution is overseen by municipal construction departments. The environmental protection administration supervises sewage disposal. The agencies are disconnected from each other, such that the department in charge of water sources does not care about water supply, and the one responsible for water quantity has no control over water quality. Each individual agency has the power of enforcement, the authority to collect fees and the right of approval. There are even two different regulations, issued at two different times, which pertain to water prices.[17]

In Beijing, Shanghai, Shenzhen and other big cities, this industrial mode of split operation has begun to change. A new water service management system and corresponding water service group which cover an overall process of proto-water supply, urban water supply and sewage disposal were established to oversee the entire industry. This unified approach to water supply requires the agencies of the state – which tend to centralize power – to limit their intervention. Moreover, these cities still lack a legal framework to support this type of public utility.

A survey of the water service industry

For a long period of time, the price of water conservancy projects for Chinese water supply has fallen short of their operating costs. Thus, infrastructure is unable to be maintained and updated, resulting in heavy losses (both financially and of water).

According to a 2002 investigation conducted by the Water Resources Ministry of over 100 major water administration units in the whole country, water conservancy projects supplied water for agricultural use at the price of only $3.61/m^3$ (1 Yuan RMB=100 Fen).[18] For industrial and household use, the price was 22.84 Fen/m^3 (USD $0.02) and 23.95 Fen/$m^3$ (USD $0.03) in 2001.[19] Water supply projects typically sell water at a price than is much lower than the cost of supplying it. Such projects are obviously not profitable.

During the past two decades, the price of water sold in most cities in China increased substantially.[20] The price increases were more dramatic in some northern cities which lack water. In 36 large and medium-sized cities across the whole country, the water price for domestic users was 0.14 Yuan/ton in 1988, and rose to 1.27 Yuan/ton in 2001, then to 1.34 Yuan/ton in 2004 (more than 9 times higher than the price in 1988). The price in Beijing has increased more quickly compared to other cities, and now exceeds 5 Yuan/ton.

However, these price increases have not changed the loss-making situation of the water supply industry. Research shows that water supply enterprises have experienced an even more dramatic increase in cost. This is largely because of depreciation costs relating to maintenance of the pipe infrastructure (Wang Xinbo 2005). In addition, the deterioration of water quality has led to an increase in both the proto-water price (mainly the water resources fee) and the actual costs of processing water.

The drainage industry faces even heavier losses, since there are generally low collection rates for the sewage disposal fee. By the end of 2003, 325 (49.2 percent) out of the 660 cities began to collect the fee.

Sewage disposal

In those cities where sewage disposal exists, it is characterized by low pricing standards, a low rate of payment and collection, and poor cost recovery in general. In one-fourth of the cities – including nine cities in the Jiangsu province – the sewage disposal fee levied for households is less than 0.3 Yuan/m (in those nine cities, the lowest fee is 0.10 Yuan/m). Among 12 cities of Guizhou province which do levy a sewage disposal fee, the prices for 11 cities range between 0.2 to 0.3 Yuan/m. In Shanxi, only 35 percent of the sewage disposal fees are collected, because loss-incurring enterprises and some disadvantaged inhabitants have continuously defaulted on their payments. The same situation applies in other areas (Construction Ministry 2005).

Huainan city (in the province of Anhui) collects 6,000,000 Yuan (approximately US $745,000) in sewage disposal fees every year. However, this constitutes only 15 percent of the city's costs to treat the sewage. Shenyang enacted a new price in September 2005. However, many civilians and enterprises are deeply discontented with it, and even appealed to the State Audit Administration (Li 2004).

Thirty-five large and medium-sized cities have all started to levy a sewage disposal fee. The fees for domestic and other uses are, respectively, 0.41 Yuan/ton and 0.58 Yuan/ton. Though the sewage disposal fee has been increased considerably, the operating cost of sewage disposal factories (without considering the costs of infrastructure) has been kept at 0.8 Yuan/ton. There is still a wide gap between the actual costs of sewage disposal and total cost recovery.

In addition, most cities lack suitable infrastructure to handle urban domestic sewage. The construction of an appropriate pipe network lags far behind demand, so some sewage disposal factories cannot collect their designated amount of sewage. Only one-third of the entire country's established sewage disposal factories operate normally.

Water prices

Public utilities set the price of water using the "cost-plus-yield rate" method. Yet the main reason behind cost inflation in water supply and discharge enterprises relates to their lack of an appropriate accounting system, which means there is no oversight of costs (Wang Xinbo 2005). At present, these enterprises have no industrial accounting system, nor any supervision over accounts. So, the overall situation in the water service industry is characterized by excessive investment, lack of maintenance and a surplus of redundant personnel. The enterprises have adopted accounting methods of government agencies, which exclude a basic tally of costs and assets.

The Chinese government formally introduced a price hearing system, in which the public can participate in price-setting. But inhabitants are generally unwilling to pay, and are quite disgusted with frequent increases in water prices.[21]

Market reform

Various government departments have released restrictions on foreign and nongovernmental capital, in order to improve the efficiency of infrastructure and public utilities, and especially to reduce the financial burden on the state. The State Planning Commission, Construction Ministry, and local and municipal governments have issued rules and regulations which encourage both foreign and domestic investors to engage in the private management of public utilities.[22]

Furthermore, the Water Resources Ministry has hastened market reforms for basic water conservancy infrastructure. A series of experimental property rights reforms were carried out for small-sized water conservancy projects in Zhejiang, Shandong, Shanxi, Anhui, Jiangxi and other places. Supported by the Agricultural Water Use Association, Ningxia carried out property rights reforms, such as contract and auction of the secondary and tertiary canals.

At present, there is no legal obstacle to non-state investment in basic infrastructure and public utilities. The 2004 "State Council's Decision on Investment System Reform" again reiterated the state's desire to attract private capital to provide public services, and to construct relevant infrastructure. It promises reasonable returns and a certain ability to recover investment.

While the laws and regulations may seem attractive, the policy environment for public utilities is actually rather uncertain. Because of this, the Chinese water supply and discharge industry has paid a high price for using foreign capital. Foreign investors were promised high returns, and as a result, some cities have acquired a substantial financial burden.

One of the cases I studied concerned a joint venture between Shenyang Water Company, a state-owned enterprise which is the largest water supply enterprise in Northeast China, and "Sino French Water" – a joint venture between Ondeo Services (France) and the Hong Kong New World Infrastructure Company.

The case involved a non-standardized contract between the contracting parties, beginning in 1996. The Chinese side of the contract was negotiated by inexperienced officials from the Shenyang municipal government (i.e. they were not representatives of the Shenyang Water Company), who were under pressure to attract foreign investment. The government officials did not involve any intermediary agency or law firms (a standard practice when negotiating international contracts). The foreign side drafted all of the legal documents. The contract involved no bidding procedures and no other businesses were offered the opportunity to invest.

The contract enabled Sino-French Water to acquire a 50 percent stake, and Shenyang Water Company would purchase all the water produced by the joint venture. The contract guaranteed a fixed minimum annual return rate of 18 percent – a rate which was much higher than the average long-term interest rate for foreign capital at the time, about 5–6 percent. The company not only benefited from a contract which guaranteed fixed returns, but also obtained its stake at a reduced cost. However, no provisions were included in the contract to guarantee the benefits of the Chinese side.

In the end, the contract was terminated because retail prices could not keep up with the costs. However, this was at a cost to the Chinese side of over RMB 300 million – 2.4 times the sum originally invested by the foreign side. Due to the high price for inviting foreign investment, plus various other reasons, Shenyang Water Company suffered a loss of as much as over RMB 200 million by 2002.

In fact, the case of Sino-French Water was not unique. At the time, it was common for local governments to attract foreign investment by providing a high fixed return. This unequal agreement – in which risk and reward were not balanced – caused great losses to many places. The Chinese government ended this kind of contract, since it stirred up many disputes between contracting parties, and also with the public (Xinbo and Shaoqi 2004).

Although the Shenyang case was costly, it helped to bring about subsequent reforms to the process of developing contracts with foreign investors. In May 2002, the French Veolia Group won an open international tender to supply water in Pudong, Shanghai, with a high bidding price of 2.1 billion Yuan (more than US $260.7 million) for a 50 percent share of Shanghai Waterworks Pudong Ltd. over an operating period of 50 years.

Evaluation

It has been difficult for China to promote market reforms in the area of public utilities. There is a weak legal environment, so oversight of the water service industry is difficult. At the same time, public utilities rely primarily on capital acquired from public finances and state-owned banks. Keen to attract a large share of those resources, many public utilities have over-estimated local demand for water, and have thus over-invested. Public utilities also lack necessary asset maintenance. At the same time, their costs are uncontrolled, because they lack the discipline that market processes provide to private sector firms.

The root cause of the problem lies in the fact that our water industry is not market-oriented; it still strongly reflects the monopolized operations which were adopted during the era of the planned economy. Public officials operate with the mentality of "low water price + loss + financial subsidy". This has led to a practical problem when public sector water enterprises are trying to become commercial operations. Specifically, they are unfamiliar with how to achieve a balance between investment, risk and a satisfactory return.

The true benefit of private-sector provision of water services is that it not only guarantees that water companies have direct contact with their customers (and thus have an incentive to provide high-quality services), but also helps water companies to perceive market demands, engage in supply-oriented production and reduce waste of resources. Unlike the past, foreign suppliers will be involved the entire course of production, infrastructure, sale and supply (*Economics Reference* 2004). Thus, foreign investment in water supply for the Chinese market hopefully entails the beginning of a new era for water services.

It is urgent to reform the present system and legal environment, to promote private investment in water services. This will require government reforms that both strengthen democratic supervision and enhance the degree of public participation, and create an effective, transparent and accountable regulatory system.

Summary and conclusion

This chapter described three aspects of China's existing water governance, and evaluated its characteristics and effects. China's serious water crisis is rooted in the present system of water governance. This system relies upon "state ownership" of water, which is carried out in the form of top-down planning and hierarchical political structures.

This top-down management creates high operating costs, which government authorities are unable to control. First, the state lacks the information which would in principle enable it to control water resources. Second, even if it did possess that information, the state lacks the ability to execute and enforce policies for water users. Thus, the idea of a "Total Quantity Control" is meaningless.

Water shortages and water pollution have exacerbated environmental problems and also threaten human health. Although a few local achievements have been made, generally the state's attempts to control pollution have not improved the situation. The environmental protection agencies can often control the mainstream and the watercourse, but they neglect tributaries and groundwater.

At the same time, academic researchers in China have not developed an adequate theoretical and practical explanation of water rights. Should private water rights really derive from state ownership? In my view, the explanation of water rights should be different from the private law origin of the public rights.

The role of the government is not to allocate water resources, but to enable bottom-up mechanisms that will do so. The current system has created artificial water scarcity, which has led to conflicts. Private ownership would enable people to trade water with each other, and thus resolve the conflicts among competing uses between households and firms, as well as between communities and the environment.

There are many advantages to systems which rely upon property rights rather than arbitrary government power. Especially because China is a country of diverse regional cultures and resource environments, a system of water rights would create advantages in terms of variety and flexibility, and enable a more extensive use of local knowledge. In combination with water user associations, regional alliances, and a legal framework which enables and respects decentralized decision-making, such a system would go a great way towards solving China's current water problems.

Although China is a large developing and transitional economy, it possesses a historical tradition of oriental despotism. A legal system which protects private property rights has never emerged in over 1000 years of civilization. One western scholar – Karl Wittfogel – observed keenly that the oriental private property was "beggar's" property (Wittfogel 1957).

Wittfogel and others have argued that water governance played an important role during the evolution of Chinese civilization. It is likely that centralized control over water helped to reinforce the despotism that pervaded China until recently.

Past dynasties all treated the public ownership of water resources as God's truth – and power over water meant political power. This tradition continues today with the state's control of water resources stipulated by the water law, from which water authorities at all levels of government derive their authority.

However, the attempt by the state to control water is at odds with the fact that households, farmers, and businesses need to use water resources. Actually, when the state claims that water is a public resource, this conceals the fact that water is shared among man and ecosystems. Even if water is a "public" resource, we should not deny private use and management of that resource. The fact that individuals and organizations compete over state-allocated water, which often leads to conflicts, shows why we need to define and protect water rights.

China is now experiencing an evolution in its management of water resources. Inevitably, it is a gradual and difficult process to move a system which relies upon centralized state power one that relies upon decentralized decision-making. Yet China's ability to achieve sustainable development depends to a very significant degree on whether or not it successfully shifts from this top-down approach, to a bottom-up approach in which water resources are allocated and managed through markets and other local institutions.

Notes

1. It is uncertain what environmental impacts will occur (especially in the southern area) as a result of the South-to-North Water Diversion Project. The Project has not encountered many objections by water rights organizations, since the Yangtze River has no clear definition of water rights. Phase 1 of the Project is under construction at present. The provinces and cities involved will share their water rights in accordance with their contributions to the Project. The Central Government's investment accounts for 40 percent of the total.
2. Using a conversion rate of US $1 to Yuan 8.05, February 2005.
3. See Table 5.2, Hou (2001, 52) for further information.
4. The 1993 "Implementing Measures for Water Abstraction License System" and the MWR "Regulation on Request, Examination and Approval Procedures of Water Abstraction License" provide regulations on the rights and obligations acquired through the abstraction license.
5. "Regulations on Management of Water Resources in Shanxi Province" drafted by Shanxi Province in 1983 was the beginning of management through the abstraction license in China. At present, there is no uniform way that water resources fees are collected; most of the regions have their own individual methods.
6. Article 12, Article 45 and Article 48 outline the details of administration and allocation of water.
7. For example, it is stipulated in Hebei province that the provincial department of water resources, municipal and county authorities, have approval powers for water abstraction of (respectively) 10 million, 1 to 10 million and below 1 million cubic meters.
8. The water resources fee is disputable in theory. Some consider it to reflect the commodity price of water resources (i.e. it demonstrates the scarce nature of water resources). Some consider it to be an expense for the management of water resources conducted by related administrative departments. The reality is that the charging standards are highly variable. Some are several cents, and some several jiao (one jiao = 1/10 of one yuan). Northwest China – the area with most water scarcity – boasts the lowest water resources fee. A major problem is that collection rates are very low, and some local governments and users dodge the payments.

Water governance in China 173

9. At the upper reach of Zhanghe River (a branch of Haihe River), a violent conflict broke out between the villages along both banks of the river. Since it was a trans-provincial dispute, the Central Government set up a special "Administration of the Upper Reach" to coordinate it. The Administration was granted relatively high powers, which were extremely costly. For related stories see Xiong (2005) and Liu (2002).
10. Wang Yahua (2005, 125) compared water control structures between western countries and China. His analysis suggests that the power hierarchy in China is from upper to lower levels, and that of western countries is from lower to upper levels. He did not analyze how the western systems evolved.
11. See Liu (2005), 266–267 for discussion.
12. See Liu (2005), 268.
13. The "Total Quantity Control" Program and the permit system for wastewater emissions, as well as implementation details, are set out in the "Law of the People's Republic of China on Prevention and Control of Water Pollution (1996)".
14. The earliest argument occurred between a well-known journalist of the Xinhua News Agency and some vice president of the Chinese Environment Programming Institution. The former rebuked the huge investment in pollution control on the Huaihe River has had no beneficial effects. The latter said that the journalist had the wrong data and had overlooked the obvious improvement in the river's water quality.
15. This section draws on Hu (2005). Hu is Chief Editor of an influential Chinese magazine called *Caijing*.
16. Some people guessed that an earthquake was coming, and tried to run away from the railway station and airport.
17. The Administrative Methods for Urban Water Supply Price (1998) and the Administrative Methods for Water Resources Project Water Supply Price (2003).
18. Though the price of agricultural water is very low, farmers actually pay a relatively high rate. Villages, towns and governments at the basic level levy various management costs on them. Shanxi, Sinkiang and other places execute a "terminal water price" system characterized by "one price to the household" and "one ticket for collection." They enforce the public demonstration system of "water

price, water quantity and water rate" which effectively reduces those intermediate links; contains unjustified price rises, irresponsible collection, water fee withholding and embezzlement, and; lightens water rate burden on farmers.
19. See Xinbo (2005), where I conducted relatively detailed research on investments and costs of Chinese water supply and discharge enterprises. There is a surplus of supply capacity among supply enterprises. At the same time, the quality of their assets is falling while leakage rates in the pipe network have been increasing year after year.
20. See Hou (2001), Table 2.12, p.22.
21. In 2002, the price of water paid by Chinese inhabitants was equivalent to 1.2 percent of the family income, much lower than the international level of 4 percent. However various public-opinion polls show that there is a very low percentage of people who support an increase in water prices. See also Hou (2001), Box 5.1, p.58.
22. In December 2001, the State Planning Commission (the present Reform and Development Commission) published *"Some Suggestions on Improvement and Introduction of Nongovernmental Investment"* which opened public utilities to nongovernmental capital, and encouraged domestic private investors to participate in private management of public utilities. Other agencies – including the Construction Ministry and the State Environmental Protection Administration – have subsequently issued similar reforms. Local governments, including Shenzhen and Beijing municipal governments, have also issued rules and regulations to comprehensively promote private management of public utilities.

References

China Economics Times (2004). "Different monitoring results with three conclusions about Huaihe River control." 23 September.

Construction Ministry (2005). 16th September, 2005, "The National City Sewage Disposal Situation Circular" by the Construction Ministry.

Economics Reference (2004). "Sudden change in the water service market: foreign capital retires quietly while social capital swarms in crazily." *Economics Reference*, 15 April.

Hou, Eve (2001). Nine Dragons, One River: The Role of Institutions in Developing Water Pricing Policy in Beijing, PRC. MA thesis, University of British Columbia, Canada. Online: http://www.chs.ubc.ca/china/fulltext.htm

Hu Shuli (2005). "Environmental protection events are matters of overall importance which must not be neglected." *CAIJING*. 24 November.

Liu, Bin (2002). "Water Rights in China." Paper presented at International working conference on water rights: Institutional options for improving water allocation. 12–15 February, Hanoi, Vietnam. Online: http://theme5.waterforfood.org/pubs/200302conf/papers/15China.pdf (visited 1 February 2006).

Liu, Bin (2005). "Institutional Design Considerations for Water Rights Development in China". In *Water Rights Reform: Lessons for Institutional Design*, Bryan Randolph Bruns, Claudia Ringler, and Ruth Meinzen-Dick (Eds.) pp.261–279. Online: http://www.ifpri.org/pubs/books/oc49.asp (visited 1 February 2006).

NPC Information Center (2005). Comments about the implementation of the 'Law of the People's Republic of China on Prevention and Control of Water Pollution' by Mao Rubo, Director. 29 June.

State Environmental Protection Administration (2004). Report On the State of the Environment in China 2004. http://www.zhb.gov.cn/english/SOE/soechina2004/water.htm

Wang Shucheng (2003). "Water resources management of the Yellow River and sustainable water development in China." *Water Policy*, vol. 5, no.4., pp.305–312.

Wang Xinbo (2005). *Research on Chinese Urban Water Price Forming, Supervision and Control*.

Wang Xinbo, He Shaoqi (2004). "Abortion of a Political Achievement –the Complicated Case about the 8th Water Factory of Shenyang," Beijing: Unirule ("Tianze") Economics Institute, August. Online: http://www.fnfmalaysia.org/article/shenyang.pdf

Wang Yahua (2005). "Shuiquan jieshi" ("Explanations of Water Rights"). Shanghai Sanlian Press.

Wittfogel, K. (1957). *Oriental Despotism: A comparative study of total power*. New Haven: Yale University Press.

Xiong Xiangyang (2005). PHD dissertation on water conflicts in the Zhanghe River.

7 The reality of water provision in urban Africa

Franklin Cudjoe and Kendra Okonski

It has been six months since Franklin rented his new apartment in Accra, Ghana's capital city, but not a single drop of water has flowed from the taps. Three months before the lease was signed, the landlord had complained to the city's water authorities. A series of visits by patrol teams, coupled with promises that the anomaly would be corrected, came to naught.

Meanwhile, the water bills continued to pour in. When Franklin finally pleaded for "reconnection", the water company insisted that he must pay the accumulated bills – for the entire nine months, during which no water was received – otherwise his apartment would be disconnected for good. He chose the latter and is now waiting for the private operators – who will soon take over management of Accra's water supply – to commence work before daring to ask for water again.

According to the World Health Organisation (WHO), 79 percent of urban Ghanaians have access to water (Table 7.1). If by "access" it refers to a pipe that leads to the house, then this figure may be accurate. But if Franklin's experience is anything to go by, such access is of little practical utility.

A study conducted by Ghana's Institute of Economic Affairs estimated that about nine million people, or 45 percent of the population of Ghana, lack adequate water supply. Of the 11 million inhabitants who do have "access" to a water system, 78 percent live strictly in urban areas. Ghana's 96 pipe-borne water systems have an installed capacity of 670,000 m^3/day (IEA Ghana, 8–9) and deliver

Table 7.1 **Population, water supply and sanitation coverage, 1990 and 2002, for six African countries**

	Year	Population			Water supply coverage (%)				Sanitation coverage (%)	
					Total		Urban			
		Total (×1000)	% urban	% rural	Total	House connections	Total	House connections	Total Total	Urban Total
Cote d'Ivoire	1990	12,505	40	60	69	24	74	52	31	52
	2002	16,365	44	56	84	33	98	65	40	61
Ghana	1990	15,277	36	64	54	14	85	35	43	54
	2002	20,471	45	55	79	24	93	50	58	74
Guinea	1990	6,122	25	75	42	10	70	37	17	27
	2002	8,359	34	66	51	8	78	23	13	25
Kenya	1990	23,585	25	75	45	22	91	58	42	49
	2002	31,540	38	62	62	29	89	56	48	56
Senegal	1990	7,345	40	60	66	22	90	50	35	52
	2002	9,855	49	51	72	40	90	71	52	70
U. Rep of Tanzania	1990	26,068	22	78	38	10	79	30	47	51
	2002	36,276	34	66	73	16	92	44	46	54

Source: WHO-UNICEF (2006)

only 570,000 m³ – about half what Ghanaians demand (*Ghanaian Chronicle* 2006).

In central Accra, a large proportion of city dwellers have access to piped water. But only 18 percent of those on the city's margin are so fortunate. Half of those who lack access to piped water depend on water vendors and 30 percent depend on streams and wells.

Among other things, these details show the inadequacies of relying on statistics provided by governments and intergovernmental bodies such as the WHO, which rely primarily on member governments for their statistics. On the one hand, governments may exaggerate the extent to which they supply people with water by defining the boundaries of the city as those areas which are supplied with water and other municipal services. On the other hand, they may under-estimate actual levels of supply because – for the same reason – they do not recognise as legitimate the informal sector entrepreneurs who supply water and sewerage services.

This is not to say that water and sewerage services are satisfactory in most African countries, or that wide scale improvements are not needed. Rather it is to say that entrepreneurs are filling the water and sanitation gap that has been created both by neglect and excessive intervention by national and municipal governments.

This chapter analyses some of the primary issues which prevent improved urban access to water across Africa. The region is often characterised as being water-deprived. Yet the main problem is not necessarily physical quantities of water, but the efficiency with which water is used, and the lack of ability to utilize more. Note that although this chapter focuses on urban water, the problems of water scarcity in many rural areas are similar or worse – however, space constraints prohibit detailed discussion of these issues.

Background

Africa is perceived to be a water-scarce region. Yet the problem is not necessarily an absolute physical scarcity of water, but a relative scarcity – one of an economic nature. Scarcity in urban areas especially relates to mismanagement by governments. In contrast, markets have been shown to be superior at addressing resource scarcity of all kinds. So why is water still allocated by governments in so many African countries?

Since the time of independence, many African nations (with a few notable exceptions) have viewed provision of water to be the responsibility of government. Motivated by paternalistic considerations and the philosophy of "need", these post-colonial African governments have presented water and sanitation, along with other community services, as basic public services to which all citizens are entitled, and for which generous subsidies (paid from government coffers) are required.

At the same time, economic growth has stagnated in many countries, while cronyism, nepotism and general ineptitude have characterized the performance of many (African) government officials.

In 2000, the WHO estimated that Africa contains 28 percent of the world's population without access to improved water supplies, and 13 percent of the world's population lacking access to improved sanitation (WHO 2000, 6.1). Only 62 percent of people in African countries have access to improved water supplies, and only 60 percent have access to sanitation (these figures include urban and rural access) (ibid.).

In these areas, public water services have typically been assigned to a single city-wide water authority. The ability of governments to deliver water has been negatively affected by a number of factors, many of the government's own making. One problem is that, whether intentionally or not, land-use planning systems are failing to recognize that a growing proportion of Africa's population has moved (and is moving) to urban areas (see WHO 2000, 4.2).

Table 7.1 illustrates, among other things, how urban populations grew between 1990 and 2002 in six representative African countries. Table 7.1 also provides the WHO's estimates of the proportion covered by water connections and sewerage ("sanitation"), for six countries in this table are discussed throughout this chapter.

Approximately 27 percent of Africa's urban population lives in dwellings on the outskirts of formally recognized "urban" areas, often referred to as shanty towns or slums (ibid.). Because these areas are not formally recognized by the government, they are denied services such as water and electricity that are (nominally at least) supplied by the government. A 1993 study showed that:

> *In most cases, land-use planning defines a city according to fully serviced areas. Those areas where low-income families live without access to water and sanitation, by this definition, are not considered urban land. Similarly, often no cadastral database exists for families living in informal peri-urban settlements. Hence, they are not included in municipal development plans.* (Solo et al. 1993, 19)

The same study outlined the following barriers to improved water and sanitation services (Solo et al, 1993, viii):

- Cities are often defined according to fully serviced areas, which do not always include the poor.
- Planning is by prohibitive zoning.
- Population growth rate is not always taken into account.
- Prohibitive land-use planning distorts the urban land market.
- City planning and building codes define housing without services as unacceptable.
- Legalization and property rights must first be approved before ownership of land is recognized.

Underlying these barriers is the fact that poorer people often are viewed with disdain by government officials and the elite. The poor lack both political clout and legal standing. As a result, they are unable to obtain public services that are ostensibly theirs by right. Moreover, they lack formal title to the land upon which they live. Because of the generally burdensome regulatory environment and high cost of enforcing contracts through the law courts, the poor tend not to participate in the "formal" economy.

What all of this means in practice is that many urban residents – especially the poor and those who reside in peripheral urban (peri-urban) areas – have insufficient access to water.

Public versus private provision of water

In many of Africa's large urban areas, municipal water systems are characterised by heavy losses – both financially and of water itself. Given a general failure by the public sector to provide reliable and adequate water at an appropriate price, some governments have engaged private sector water providers to manage the supply.

The broad aims, which have varied in each context, were to increase efficiency, and to generate enough revenue to cover operating costs and thus encourage investment. The arrangements in such situations also varied: in some cases ownership has been transferred to the private company. In others, a lease or a management contract is agreed, under which a company oversees existing infrastructure, ownership of which is retained by governments. The brief case studies below are intended to provide examples of the situation pertaining to urban water systems in African countries.

Dar es Salaam, Tanzania

A 1997/98 study found that the government-managed water system of Dar es Salaam, population approximately three million, had 98,000 connections. In addition, loss rates were horrific: 53 percent of water was "unaccounted-for." Meanwhile, although 88 percent of water production was billed to consumers, only 54 percent of this was paid.[1]

Low income areas were "essentially un-serviced by utilities." Instead, "third party initiatives" filled the gap created by government. These include, "vendor-supplied water services on push-carts, shallow wells belonging to more financially influential residents, and to a limited extent utility water connections in a few residences"

(Wandera 2000, 15). Where such water connections exist, "these residences sell water to neighbors and to vendors who subsequently deliver water to areas of shortage." Sanitation services are also provided by such third party initiatives, "typically by way of traditional pit latrines that are emptied by frog-men" (ibid.).

Nairobi, Kenya

A 2005 study of government water supply to Nairobi, population three to four million people, shows that the city has 182,295 legal water connections (Gulyani et al. 2005, 4). The system loses approximately 50 percent of its water, including both "technical losses" (i.e. leakages) and "commercial losses" (i.e. unbilled and uncollected revenues and theft). However, the system itself has unreliable data on household water use and losses by households and other entitites, and it has insufficient metering. Thus, "bills are based on presumed consumption. The billings system is poor, collection efficiency (or revenues collected as a proportion of total billed) is 65 percent, and accounts receivable stand at more than two years of billings" (ibid.).

Conakry, Guinea

In 1989, Guinea extended a private lease agreement for operation of water services to 17 cities and towns, including the capital city of Conakry.

Before the 1989 reforms in Conakry's water sector, access to water in the city was estimated at less than 40 percent (Menard and Clarke 2000, 1). At that time, the city's then population of around one million people was serviced by about 3,100 legal unbilled connections, 10,200 unbilled connections and an (estimated) 4,000 illegal connections (ibid.). Unaccounted-for water was at least 60% in 1983 (ibid. 10).

The pre-reform water system was "almost inoperative," yet it employed a staggering 42 people per 1000 customers (Noll et al. 2000, 25). At the same time, "tariffs did not cover operating costs, much less capital costs" (ibid.) The system was "on the verge of collapse, delivering poor quality water (in some cases through lead pipes) for only a few hours a day" (ibid. 21). Owing to a lack of safe water and sanitation, the city had high infant death rates and frequent outbreaks of cholera (ibid.).

Following the 1989 reforms, "Capacity more than doubled, water quality and service improved dramatically, the population served almost doubled, and coverage expanded from 38 to 45 percent" (ibid. 37). Nevertheless, for several reasons, it seems that weak institutions have created difficulties for the private operator and weakened its ability to deliver the expected benefits: "Guinea's weak institutions made it hard for the government to negotiate and commit to an affordable water tariff with the private operator. These same weaknesses paralyzed operation under public ownership" (ibid.).

The main institutional failures were:

1. "The leaseholder's main risk has been from confiscation of quasi-rents by government and consumers through non-payment of water bills in a legal system where cut-offs are hard to enforce" (ibid). "Interestingly, rents are not expropriated in Abidjan [discussed below] or Conakry through low consumer prices, but by forcing the operator to supply non-payers, especially the government" (ibid. 39). The government has been "the worst source of accounts receivables" paying only 10 percent of its bill in 1993 (ibid. 31).[2]
2. Despite widespread metering, Guinea's laws "make it hard to disconnect non-payers permanently or penalize persons for illegal connections" (ibid, 31). In 1996, "58 percent of private bills went unpaid" (Cowen 1999).
3. Unaccounted for water (UFW) – an "indicator of the efficiency of a water utility" (Noll et al. 2000, 12) is still very high. "UFW was not accurately measured before reform and remained very high at 50 percent in 1996, because of the inability to collect bills and prevent illegal connections" (ibid. 32).

One reason that Conakry has high water tariffs is "the costs of the system sized for a larger customer base are spread over so few paying connections" (ibid. 34). In 1996, the city had 17,638 legal connections, and 130 stand pipes (estimated to serve 975 people each) (ibid. 31–32). "Conakry's tariff was set to cover the cost of servicing the debt contracted prior to the lease, even though much of this was inefficiently invested. Thus, Conakry has a system with abundant, low-cost water that is priced beyond the means of many citizens" (ibid. 32).

Abidjan, Cote d'Ivoire

Cote d'Ivoire has the longest running private water system on the continent, operated today by the French multinational company, SODECI. Starting in 1959, Abidjan's water system was managed with a lease contract (ibid. 18). In 1986, the private operator had only 9.8 employees per 1000 connections – which is a small number compared to public sector water systems in the region (Menard and Clarke 2000, 9).

Abidjan (population three million) has a water system that uses a "cost plus pricing regime which passes all costs on to consumers" (Noll et al. 2000, 30). "Metering, billing, and collecting from private consumers are almost universal; the main exception has been large accounts run up by the government" (ibid). As a result, the billing recovery rate is about 80 percent (Lauria et al. 2005, 27). In general, "large volume consumers, who are almost all industrial, cross-subsidize all other users, including government, while small consumers (less than 18 m^3) pay least, about a third of the tariff on large volumes (over 300 m^3). Consumers in the rural areas within the district are also cross-subsidized" (Noll et al. 2000, 30–31).

Dakar, Senegal

In 1996, Senegal's water sector was reformed. A public holding company – SONES – is under contract to the country's Ministry of Water to provide water services to the people of Senegal. SONES holds and owns assets, and then leases operating services to SDE, a private enterprise. The effect of this transfer from public to the private sector management has been dramatic. Between 1996 and 2001, the number of clients increased by 35 percent (from 241,671 in 1996 to 327,501 in 2001) (Brocklehurst and Janssens 2004, 21).

By 2000, SDE was supplying 85 percent of households in Dakar, a city with a population of 2.5 million (Collignon and Vézina 2000, 16). Between 1995 and 2002, "private water connections increased from 135,414 to 181,824", a 34 percent increase which exceeded the planned target. Likewise, the number of public standpipes increased by 50 percent (from 940 in 1995 to 1424 in 2002) (Brocklehurst and Janssens 2004, 21). Meanwhile, the billing recovery rate "is more than 90 percent in Senegal" (Lauria et al. 27).

Since 1996, many areas which were previously un-served have been connected to the network, as the system was expanded and extended by SDE, SONES and by private third parties. "Private developers have paid for and constructed more than 50km of water distribution network every year over the last three years, or 60 percent of additions to the network" (Collignon and Vézina 2000, 13). "The entire network increased [by 23 percent] from a length of 4319 kilometers in 1996 to 5330 kilometers in 2001" (Brocklehurst and Janssens 2004, 22).

Although agencies of government – especially municipal governments – initially were averse to paying their own bills, the government implemented "corrective measures to reduce the high water usage of public sector clients, budget annual public agency consumption, and pay government water bills within two months of their being served." As a result, "Senegal is one of the few countries in the region in which the government does, eventually, pay its bills" (ibid.).

Discussion

This small sampling of urban water systems in several African countries enables us to identify some of the unifying characteristics of public sector water provision. These characteristics are provided in context of economist Gabriel Roth's observations (1989) about public sector services:

Public provision	**Private provision**
High levels of water loss and insufficient metering.	Lower levels of water loss and widespread metering.
A lack of competition.	Varying degrees of competition.
Prices that do not reflect the actual cost of water processing and delivery.[3]	Prices that reflect operational costs plus the opportunity costs of capital.
Heavy financial losses, but a poor ability to recover costs and, thus, little if any ability to fund future infrastructure.	Profitable enterprises that have the capacity and incentive to expand services both geographically and to the existing customer base.
Low transparency and accountability to customers/recipients of the service.	More transparent; operate on the basis of contractual relationships with customers.
Rent-seeking by public officials.	Rent seeking limited by obligations to abide by contractually agreed terms.
An inability to cope with a dynamic urban situation, in particular, a growing urban population.[4]	Dynamically responsive to changing demands – expanding markets seen as an opportunity, not a threat.

These conclusions concur with Roth's analysis regarding public sector enterprises:

- First, governments tend to make decisions concerning public enterprises based on political considerations, granting favors to one interest group at the expense, and to the detriment, of another interest group or of society as a whole.
- Second, because such enterprises are in fact government monopolies, they suffer many of the same problems attributed

The reality of water provision in urban Africa 187

to private sector monopolies. Due to lack of competition, government enterprises are prone to inefficiency. Employees have insufficient incentive to provide the best service possible to customers.
◆ Third, lack of competition often leads to huge financial losses, which taxpayers are forced to cover. The need of governments in poorer countries to cover billions of dollars in such losses from their budgets has diverted enormous resources from other crucial social needs and contributed to huge debts.
◆ Fourth, because taxpayers ultimately cover such losses, and because of the considerable political power of workers employed in government enterprises, such enterprises rarely shut down even when they fail to meet public demands.
◆ Fifth, government economic forecasts are usually based on relatively few "scenarios" or projections of demand and are divorced from market processes. Thus government management of any given industry tends to be based on incomplete or outdated information. (Roth 1987, 4–5)

In contrast to public sector and government-run water systems, the private sector has performed substantially better – even when it has merely been responsible for managing the system.

The right kind of "privatization"

Opponents of private water ownership and management claim that access to water is "a human right" and that water itself is a "common good." They specifically oppose the sale of water to multinational corporations. The following is a typical example:

Water belongs to the Earth and all species and must not be treated as a private commodity to be bought, sold and traded for profit. Because the global water supply is a shared legacy, protecting it is a collective responsibility – not the responsibility of a few shareholders.[5]

These opponents claim that privatization is the process by which public utilities are sold to large private companies, or are given leases to supply water through existing infrastructure. While this is one way in which water has been privatized it is by no means the only way and it is often not the best way.

If governments do choose to sell existing infrastructure to private companies or to extend management leases, it is important that this process occurs in an open and transparent manner – for example through an open auction.

Failure to do so often leads to cronyism, with government monopolies being transferred to members of the ruling clique or to a foreign corporation that has paid a bribe to public officials. The company may retain the monopoly over service, and it is plausible that this may increase the price of water such that the poor cannot afford it. A foreign-owned company will tend to repatriate its profits, to the benefit of shareholders which typically are located elsewhere.

But to criticise private ownership and management of water on the basis of such corrupt transfers is hardly fair. In fact, the criticism should more fairly be directed at the government which pushed through the "privatization." Moreover, these problems will in most cases have existed long before the privatization, and continue to plague not only the urban water systems but economic development in the country more generally.

This is not to excuse poor behaviour in terms of contract negotiations or the other complex procedures involved when businesses deal with government. However, it is important to recognize that privatization does not occur in a vacuum; privatization itself will not overcome a poor institutional setting in which governments are unaccountable, corrupt and inept.

Opponents of privatization continue to claim that governments can supply adequate, reliable high-quality water. However, on the basis of the small but representative sample of countries and cities considered in this chapter, this is incredibly naïve or disingenuous. When water is under public management, the poor generally do not receive high-quality, reliable supplies of water (at least, not from the government suppliers). Residents of informal settlements on the periphery of urban areas typically lack formal title to their land and are excluded from ever receiving such water (Collignon and Vézina 2000, 43; Lauria et al. 2005).

While some connections may receive government subsidies, these tend to be households that are moderately wealthy (ibid.). The poor often do not benefit from government-subsidized water at all. For instance, although Nairobi's government water kiosks receive subsidized prices from the municipal utility,

> *[The kiosks] are neither providing the quality of service desired by users nor achieving the utilities' objective of subsidizing costs to poor households...[The] kiosk owners charge, on average, a price that is 18 times higher than the subsidized prices at which they receive water from the utility* (Gulyani et al. 2005, 26).

Based on evidence from Côte d'Ivoire and Senegal, Lauria et al. (2005) conclude that in terms of targeting the poor, "subsidizing connections is probably better than subsidizing consumption." This is because "if connections are subsidized, the users will pay for consumption" (27).

Moreover, the opponents of private water provision seemingly fail to recognize, or ignore, that in the absence of state-provided water, the poor in many if not all African countries are in fact paying for their water. They buy it from informal entrepreneurs, who supply water and sanitation services to their fellow citizens because they make a profit from so doing. Though this "profit" is unlikely to amount to millions of dollars, the fact that poor individuals buy and sell water and sanitation services from each other demonstrates that a market does exist. The profits of these mostly one-man operations "are largely reinvested in the water or sanitation business or in other local economic activities" (Collignon and Vézina 2000, 15).

One likely reason that poor consumers purchase water from these informal vendors is that they can use their time more efficiently in other pursuits. Likewise, the vendors have specialized to fill a particular niche which was previously untapped.

The right kind of privatization should acknowledge the informal economic activity which – at least according to one quite comprehensive study ("Independent Water and Sanitation Providers in African Cities: Full Report of a Ten-Country Study" by Bernard Collignon and Marc Vézina) – is already occurring in many urban areas of Africa today (discussed below). Likewise, it should enable that informal activity to become formalised. As noted elsewhere in this volume (Robinson, chapter 8), privatization is a necessary but not sufficient condition for achieving better allocation of water.

True private provision of any good, including water, means that the private sector and the private sector alone – which encompasses entrepreneurs of all sizes, and not just multinational corporations – is able to satisfy the demand for that good. Entrepreneurs play a key role in supplying those goods and services. They identify an opportunity to solve some form of scarcity – such as supplying water to fellow residents of a slum. They seize upon that knowledge and use it to provide goods and services to customers in exchange for payment. At a very basic level, such as that observed in the informal economy of African cities, such specialization in economic activity is the key to economic development.

When buyers and sellers strike a deal in a market, "The resulting contracts convey much information to all concerned" (Kasper 2005, 8). This is the genesis of market competition: "Profitable pioneers thus find imitators, who either copy the technology (diffusion) or improve on it. Further potential entrepreneurs are informed as to what inputs can best be bought where and when, and so on. Other buyers may be inspired to obtain the product, too, and they also find where best to shop for it" (ibid.).

Thus, the word "entrepreneur" does not refer to the size or the scale of the business, but rather, to the nature of its role in coordinating the use of resources in a society. Economist Israel Kirzner refers to this as a "competitive entrepreneurial discovery process" (Kirzner 1984, 416). An entrepreneur fulfils a complex role in a market and as economist Wolfgang Kasper observes, no one should take the entrepreneurial discovery process for granted:

> *[The entrepreneur] must discover where to obtain the raw materials and components, how to assemble them in a cost-effective manner, how to distribute and service the product, how to train workers in appropriate skills, and how to finance all of these costly efforts. And he constantly faces the threat that other entrepreneurs will enter the market, and take his market share and profits. This implies a wide-ranging and complex knowledge search about conditions in an interactive and ceaselessly changing world* (Kasper 2005, 6).

The phenomenon by which entrepreneurs provide goods and services demanded by consumers is often referred to as the "market process," which is characterised by both the entrepreneurs who make it happen and, equally importantly, the environment in which those entrepreneurs are able to operate.

Many governments around the world have failed abjectly to create an enabling environment for entrepreneurial activity – of all sizes, shapes and forms – to take place. The following is an example of some of the obstacles in Ghana which prevent entrepreneurs from conducting business in a manner considered "legal" by the government:

- Entrepreneurs can expect to go through 12 steps to launch a business over 81 days on average, at a cost equal to 78.6% of gross national income (GNI) per capita. They must deposit at least 27.9% of GNI per capita in a bank to obtain a business registration number.
- It takes 16 steps and 127 days to complete the process [of complying with licensing and permit requirements for ongoing operations], and costs 1,549.7% of income per capita.
- It takes 7 steps and 382 days to register property. The cost to register property is 3.7% of overall property value.

- For a medium-size company to pay taxes in Ghana, entrepreneurs "must make 35 payments, spend 304 hours, and pay 45.3% of gross profit."
- It takes 23 steps and 200 days to enforce contracts. The cost of enforcing contracts is 14.4% of the debt. (Doing Business 2006)

Economists refer to these as "transaction costs"; they are essentially the costs of "doing business" in a setting in which property and contracts are protected by the state. To contrast with Ghana, in New Zealand it takes two procedures and 12 days to establish a business and costs of 0.2% of per capita GNI, with no bank deposit necessary.

Meanwhile, obtaining a license takes 7 procedures and 65 days (actually relatively long compared to the OECD average of 14 days) and costs 29 percent of GNI per capita. It takes two days and two procedures to register property and costs 0.1 percent of the value of the property. For a medium sized company in New Zealand, paying taxes requires 7 procedures and takes 70 hours (though companies are taxed at 44.2 percent of gross profit, which is similar to Ghana). Meanwhile, it takes 19 procedures and 50 days to enforce a contract, with the cost being approximately 4.8 percent of the debt owed.

When the transaction costs are as high as they are in Ghana, they clearly put entrepreneurs off doing business in the formal economy. This does not mean that Ghana and other countries in a similar situation lack transactions. What it means is that many or even most transactions take place in the "underground" or informal economy.

Why are formalized institutions so fundamentally important in reducing the costs of transacting in markets? Kasper explains that institutions "create a degree of certainty for entrepreneurs, which enables them to concentrate on the technical and commercial knowledge search of innovation." Likewise, these institutions "allow entrepreneurs to economise on searching for many types of knowledge about potential risks in that they establish a trustworthy legal and social environment. Thus, they facilitate the entrepreneurial mobilisation of capital, labour, technology, skills and natural resources" (Kasper 2005, 23).

Economist Gabriel Roth explains that markets deliver superior results because "the private sector, taken as a whole, has an excellent information base" which helps to determine the real cost of goods and services through prices. Private provision of services introduces competition "and thus avoids many of the problems of government monopolies". This is especially true in terms of the quality and cost: "Services must meet customer demands and therefore tend to be of appropriate quality. And production tends to be in the most cost-effective manner" (Roth 1989, 5; for more in-depth explanation see Morriss, this volume, Chapter 2; Robinson, this volume, Chapter 8).

Africa's informal water entrepreneurs

As noted above, high transaction costs discourage businesses from operating in the formal economy. In the context of water services, the public sector in many African countries has failed to provide all sectors of society with high quality water at a low cost. The state then compounds this problem by imposing all manner of restrictions against private provision, ensuring that where the poor do get water it is being supplied by informal companies at a relatively high price.

As noted by Bernard Collignon and Marc Vézina (2000, 39) in their study of water and sanitation in ten African countries, "the vagaries of unpredictable political and economic conditions found in most countries" are the foremost concern of any African entrepreneur. Most of these businesses involved with water and sanitation services they studies "remained in the informal sector." That is, they are not registered with government authorities because of government barriers to doing business.

The evidence collected by Collignon and Vézina suggests that a plethora of entrepreneurs operating completely in the private sector – albeit in the informal, extra-legal economy – are getting around the water scarcity and dearth of sanitation services which have been created by their governments. For water services, "independent providers are dominant in six of the ten cities studied and play a major role in the others, serving most of the low-income areas in all cases."

This is particularly important with sanitation services. As noted by the study's authors, "Most households in African cities – 70 to 90 percent overall, and virtually all poor households – deal with their own waste by building their own latrines or septic tanks or hiring others to do it." The public sector is "generally not involved in this area, [so] private providers dominate the market and offer services tailored to customers' needs and incomes, for the tasks that households choose not carry out themselves: masons who build latrines, manual latrine pit cleaners, suction truck operators for septic tanks, and manual or mechanized drain and latrine ditch cleaning services" (Collignon and Vézina 2000, 24).

In all ten of the cities studied by Collignon and Vézina, "manual latrine cleaners and suction truckers are well organized and widespread" (33). Private toilet operators are successful in five of ten cities.

Entrepreneurs in the informal sector are familiar with their customer base, since they are providing services to people who live around them in a similar situation – in the slums of African cities. They are likely to live and operate in "illegal" dwellings – whether slums or shanty towns. Their jobs are far from glamorous, and hard work. Because generally they are unable to acquire loans from banks and financial institutions (for example, because they lack formal title to their land), they seek loans from family members when they want to start a business.

A characterization of such an individual,

> *shows a versatile man, risk and publicity averse; capable of raising important sums of money when necessary, but without a logo or a front office. He seeks no loans from the bank, nor does he pay the city business tax, if he can avoid it. He can and does cover many bases, depending on what is most profitable today. His relations with other providers are opportunistic, governed by the practical advantage conferred, with little inclination (at least*

so far) to control or restrict the free operation of market forces
(Collignon and Vézina 2000, 37).

Most of these entrepreneurs are individuals, but some have even become medium-sized businesses. Their individual revenues may not amount to millions of dollars – but the scale of these transactions does not matter. The fundamental point is that poor people pay those entrepreneurs to provide water and sanitation in slums where they live together.

The evidence suggests that the informal sector is more responsive to demands of individuals – and sometimes it has been so successful that government kiosks have been put "out of business", as it were. An interesting case study of Mtwara, a coastal city of approximately 123,000 people in Tanzania (Wandera 2000), revealed that the municipal water utilities began with good intentions to run government-sponsored water kiosks. However, these often resulted in poor service – for a variety of reasons. One of those was that the kiosk operators were not investors in the kiosks, nor were they paid a sufficient wage for their operation services. For consumers of water, even if this water was in principle "cheaper", it was often not available at the right time and/or place due to erratic service from the utility.

So in place of poor service offered by government kiosks, private water connection holders "installed holding tanks...and offer[ed] more reliable retail services." Specifically, "the private water sellers are able to cushion themselves from the utility's service interruptions by using the water in holding tanks as buffer supplies." Where such private connections exist, "the [government] kiosks...are closed due to lack of market" (Wandera 2000, 40).[6]

A similar case study of Arusha, Tanzania, concluded that third-party interference with water retailers should be limited to regulatory price measures, and even this "must be kept within the constraints of the prevailing market forces." This is based on the author's conclusion that water retailers are "adversely affected by non-commercial institutional impositions" (ibid. 31).

From their survey in urban areas of Kenya, Gulyani et al. (2005)

conclude that those households "are acting largely as informed and rational economic decision-makers". Moreover, "there is a well-established private market for water," and that it is "similar in nature to the market for a consumption good – it works on price and quantity, all else being equal." Survey respondents, regardless of age, wealth or gender, "understand the value of water, treat it as an economic good, and have moved away from any erstwhile notions that it will be available from a public source at low or no cost" (28).

Conclusion

The poor in African countries generally lack reliable supplies of clean, piped public water as well as sanitation services. But this chapter has demonstrated that their demand for those services is being met in part by informal entrepreneurs. They view water and sanitation services as a commercial opportunity, unlike the government's approach which tends to view these extra people as a burden on existing systems.

It is a truly remarkable feat that informal entrepreneurs and their business activities are able to fill a gap created by public sector water and sanitation systems. They supply a large proportion of water, and 70 to 90 percent of sanitation services to the low-income residents of at least 10 African cities.

Sadly, many governments across Africa seem to be averse to "bottom-up", decentralized solutions to the supply of water and sanitation. As a result, these informal entrepreneurs are often castigated by government officials and the ruling elite. This helps explain why governments fail to recognize that their cities are growing. The new urban areas deserve to be legally recognized – especially in light of the fact that Africa's urbanization is projected to continue for the next few decades.

The evidence considered in this chapter seems to refute the conventional wisdom about water and sanitation, which normally suggests that the poor simply lack water and sanitation altogether.

The reality of water provision in urban Africa 197

Most importantly, it shows that governments can undertake simple policy reforms to enable those activities to occur, without creating further demands on government coffers.

If governments were to focus on reforming economic and institutional conditions more generally (a strategy which would have widespread economic and social benefits), then it is likely that decentralized entrepreneurial activity could offer an enormous potential to solve existing problems. As Collignon and Vézina (2000) observe:

> *Supporting independent providers is thus perfectly in tune with current institutional and economic trends in Africa, and it does not imply a choice between city-wide entities and independent operators. The central and municipal governments' roles are rather to see that these two kinds of providers complement each other in the marketplace and that fair competition is encouraged. Given the choice, users can be trusted to judge for themselves where to take their business.* (60)

While public sector water systems often tend to view extra people as a burden, entrepreneurs view those extra people as an opportunity. By creating an enabling environment for entrepreneurship – such as removing artificial barriers to legal business ownership, and ensuring the application of the rule of law – entrepreneurs would be empowered to provide all levels of service, small or large.

Reforming land tenure systems – especially by enabling the poor to own their property in areas which are considered marginal, such as slums and shanty towns – would also greatly improve the situation. African governments must undertake reforms that broaden economic opportunities in general. This is the way to enable ordinary Africans to have reliable, affordable, high quality supplies of the most basic necessity of life: water.

Notes

1. Based on figures from the Tanzanian Ministry of Water, cited in Wandera (2000), Table 5, p. 14.
2. Noll et al. (2000, 31) note that "the introduction of meters drastically reduced government consumption and the intervention of the French aid agency in 1996 helped the government and the company settle their cross debts and keep down arrears."
3. In the absence of market competition and the market process, it is impossible to say whether these prices are 'too high' or 'too low'.
4. Between 1990 and 2002, all of the sample African countries depicted in Table 7.1 experienced a growth in their urban populations and similarly, a decline in rural populations.
5. A quote from the website of one of the organizations which campaigns against privatization and business more generally. Online: http://www.citizen.org/cmep/Water/activist/articles.cfm?ID=9589

References

Brocklehurst, Clarissa and Jan G. Janssens (2004). "Innovative Contracts, Sound Relationships: Urban Water Sector Reform in Senegal." Water Supply and Sanitation Sector Board Discussion Paper Series, No.1 (January). Online: http://www.worldbank.org/html/fpd/water/pdf/WSS_Senegal.pdf. Cited 14 February 2006.

Collignon, Bernard and Marc Vézina (2000). "Independent Water and Sanitation Providers in African Cities: Full Report of a Ten-Country Study." Washington, DC: UNDP-World Bank Water and Sanitation Program. Online: http://www.wsp.org/publications/af_providers.pdf. Cited 14 February 2006.

Cowen , Penelope Brook (1999). "Lessons from the Guinea Water Lease." Public Policy for the Private Sector, No.78 (April). Online: http://rru.worldbank.org/Documents/PublicPolicyJournal/078cowen.pdf Cited 14 February 2006.

Doing Business (2006). Online data query: http://www.doingbusiness.org/. Cited 14 February 2006.

Gulyani, Sumila, Debabrata Talukdar, and R. Mukami Kariuki (2005). "Water for the Urban Poor: Water Markets, Household Demand, and Service Preferences in Kenya." Water Supply and Sanitation Sector Board Discussion Paper Series, No.5 (January). Washington, DC: World

Bank. Online: http://www.worldbank.org/html/fpd/water/pdf/WSS_UrbanPoor.pdf. Cited 14 February 2006.

IEA Ghana (no date). "To Privatise or Not to Privatise: The Arguments For and Against Private Participation in Water Supply in Urban Ghana." Monograph No.4. Accra: Institute of Economics Affairs.

Ghanaian Chronicle (2006). "Pianim's Plea: Let's Conserve Water." January 23. Online: http://allafrica.com/stories/printable/200601231030.html. Cited 14 February 2006.

Kasper, Wolfgang (2005). "Economics 101 for Kyoto Fans." In: *Carrots, Sticks and Climate Change*. London: International Policy Press. Online: http://sdnetwork.net/page.php?instructions=page&page_id=550&nav_id=131. Cited 14 February 2006.

Kirzner, Israel M. (1984). "Economic planning and the knowledge problem." *Cato Journal*,Vol. 4, No.2 (Fall).

Lauria, Donald T., Omar S. Hopkins, Sylvie Debomy (2005). "Pro-poor subsidies for water connections in West Africa." Water and Sanitation Supply Working Notes, No.3 (January). Online: http://iris37.worldbank.org/domdoc/PRD/Other/PRDDContainer.nsf/All+Documents/85256D2400766CC78525700600671343/$File/WSSWN3Subsidies.pdf. Cited 14 February 2006.

Menard, Claude and Clarke, George R. G. (2000). "Reforming the Water Supply in Abidjan, Cote d'Ivoire: A Mild Reform in a Turbulent Environment" (June). World Bank Policy Research Working Paper No. 2377. Online: http://ssrn.com/abstract=630747. Cited 14 February 2006.

Noll, Roger, Mary M. Shirley, and Simon Cowan (2000). "Reforming urban water systems in developing countries." Discussion Paper No.99-32. Stanford, CA: Stanford Institute for Economic Policy Research. Online: http://siepr.stanford.edu/papers/pdf/99-32.pdf. Cited 14 February 2006.

Roth, Gabriel (1989). "Bringing efficiency to the third world through private provision of public services." Washington, DC: Heritage Foundation. Online: http://www.heritage.org/Research/TradeandForeignAid/upload/87930_1.pdf. Cited 14 February 2006.

Solo, Tova María, Eduardo Perez and Steven Joyce (1993). "Constraints in providing water and sanitation services to the urban poor." Technical Report No. 85 (March), Water and Sanitation for Health Project. Washington, DC: US Agency for International Development. Online: http://pdf.dec.org/pdf_docs/PNABN953.pdf. Cited 14 February 2006.

Wandera, Bill. (2000). Tanzania Case Study: Strengthening the Capacity of Water Utilities to Deliver Water and Sanitation Services, Environmental Health and Hygiene Education to Low Income Urban Communities. Dar es Salaam Water and Sewerage Authority. Online: http://web.mit.edu/urbanupgrading/waterandsanitation/resources/examples-pdf/CaseStdyTanzania.pdf . Cited 14 February 2006.

WHO (2000). Global Water Supply and Sanitation Assessment Report. Online: http://www.who.int/docstore/water_sanitation_health/Globassessment/GlobalTOC.htm. Cited 14 February 2006.

WHO/UNICEF (2006). Joint Monitoring Programme. Website data query. Online: http://www.wssinfo.org/. Cited 14 February 2006.

8 How not to reorganise an industry: Privatisation, liberalisation and Scottish water

Colin Robinson[1]

Any industry requires two key elements if it is to operate efficiently and to serve its customers well. Experience shows that both are necessary in the 'network utilities' such as gas, electricity and water, as well as in other parts of the economy.

First, the market for the industry's products should be competitive. No supplier should have captive customers: anyone who is dissatisfied with his or her current supplier should have the ability to switch to alternative suppliers. In those circumstances, all suppliers are kept on their toes. They have powerful incentives not just to control current costs but to innovate in terms of prices and quality, so as to maintain or improve their future market positions. Moreover, because of market rivalry, cost reductions and the benefits of innovation tend to be passed on to consumers.

Second, companies in that industry should be held privately. If they are, competition in capital markets will mean that the company's shareholders will press its managers to perform efficiently, reinforcing the pressures from competitive product markets. Managers who do not respond risk losing their positions, because a falling share price will make the company a takeover target and a new management team may take their place.

In practice, markets – like all human institutions – work 'imperfectly'. Nevertheless, experiments with other forms of economic organisation – in particular establishing state corporations with monopolies of the products they sell in the belief that they will pursue the 'public interest' – demonstrate that such alternatives are inherently inferior to competitive markets with private ownership. Such enterprises have few, if any, incentives to operate efficiently, to innovate and to respond to the wishes of consumers. Managers cannot perform well in such circumstances, not necessarily because of the failings of individuals but because the system is at fault. Both economic principles and the lessons of history indicate that monopoly state corporations are doomed to failure.

This chapter examines the difficulties inherent in nationalised ownership of industries that produce goods and services for sale, discusses the problems specific to the Scottish water industry, compares these to the situation in England and Wales, and makes some proposals for improvement.[2]

Scotland's water

The previous observations about state corporations – gleaned from both theoretical observations and practical experience in Britain from the 1950s to 1980s – do not appear to have reached the authors of Scotland's water supply regime. When the Scottish water industry was reorganised in 2002, state ownership was maintained and its monopoly characteristics were reinforced. Three regional water companies, in east, west and north Scotland, were amalgamated to form Scottish Water (SW), a nationalised monopoly with a regulator, the Water Industry Commissioner for Scotland (WICS).[3]

SW is one of the Britain's larger companies in the water industry, with a turnover of about £1 billion a year and about 4000 employees. Because it is nationalised, it cannot be compared in terms of market value with the privatised water companies in England and Wales. However, Table 8.1 (which uses various physical size indicators to compare SW with the ten regional water and wastewater companies in England and Wales) shows that SW is bigger than any of the English and Welsh companies in terms of length of mains and

Table 8.1 **Scottish Water relative to water companies in England and Wales**

	Scottish Water	Ranking relative to water and wastewater companies in E & W
Length of water mains (km)	46,508	1
Length of main per property (m)	18.74	5
Length of sewers (km)	44,854	3
Length of sewer per property (m)	13.34	7
Number of water treatment works	371	1
Number of wastewater treatment works	616	4

Source: WICS (2004c)

number of water treatment works, and is ranked third by length of sewers.

Given the nature of the 2002 reorganisation, it is hardly surprising that it failed to stem the flow of public complaints about the industry. Criticism is rife in the Scottish media about SW's alleged inefficiency, its charges, its standards of service and more generally its apparent lack of concern for the interests of its consumers. New legislation passed by the Scottish Parliament in February 2005 – the Water Services etc (Scotland) Act[4] – made some significant changes to the industry and, in particular the way it is regulated, but the legislation does not address fundamental issues about the industry's ownership and structure.

Scottish Water: efficiency and standards of service

Some of the strongest criticisms on the state of the Scottish water industry have come from its 'economic' regulator, Alan Sutherland, who was the Water Industry Commissioner for Scotland before the 2005 Act. In a November 2004 report, WICS made an 'overall performance assessment' across a range of services and found that between 2002 and 2003, SW's standards of service were far worse than those in England and Wales (WICS 2004b, chapter 6). He ranked

SW's performance at a level of only 38 per cent of the worst-performing water company in England and Wales.

The comparison may seem unfair since SW was only formed in 2002 from its three predecessor bodies, and so had little time to reorganise. Moreover, SW has argued that the England and Wales industry surged ahead because of its big investment programme since privatisation (*Scotsman* 2004). However, as WICS points out, the 'asset bases either side of the border appear to have many similarities' and, in the last twenty years, investment per connected property in Scotland has matched that in England and Wales.

WICS therefore concluded that it is inefficiency in investment, not lack of investment funds, which distinguishes the Scottish water industry from its counterpart in England and Wales (WICS 2004c, 3–4). In his words, 'Customers in Scotland have paid for, and so deserve, an equivalent standard of service to that which customers in England and Wales receive' (ibid, 4.).

As explained below, efficiency comparisons between public- and private-sector water companies are fraught with difficulties, so there is room for argument about the size of the difference between SW and the companies in England and Wales. Nevertheless, there seems to be a significant lag in performance in Scotland.

Nationalisation may not be the only reason for this poor performance, but those familiar with the history of Britain's nationalised industries will find a familiar ring in the criticisms made of SW. They echo criticisms made of the 'public' corporations that were established just after World War II as part of Prime Minister Clement Attlee's nationalisation programme. Most of these corporations – including 'utilities' such as gas, water, electricity and telecommunications – subsequently were privatised in the 1980s and early 1990s. A few remain nationalised today, including the Post Office as well as Scottish Water.

Describing these nationalised corporations as 'public' bodies is misleading. 'National ownership' did not mean that the corporations were genuinely accountable to the public. Indeed, one of the main reasons for frustration with their performance was – and still is where such corporations remain – that the general public has no control over the activities of these organisations, and feels powerless to influence their behaviour. As explained below, such discontent is not so much the fault of particular individuals in particular organisations: it is an innate characteristic of a regime where nationalised corporations monopolise 'key' industries.

Seen in this light, one of the underlying reasons for the failings of Scottish Water becomes clearer: it is the result of political failure to establish appropriate ownership and structure, and an appropriate regulatory regime for the water industry in Scotland. Politicians created a nationalised monopoly, protected by statute from competition, in circumstances where such a form of organisation is ill-suited to the circumstances of the industry.

To consider remedies, we need to look to the past to evaluate why nationalised monopolies have so failed and why, in most British industries, state corporations no longer exist.

Nationalisation and its problems

The record of nationalisation

Nationalisation began with high hopes in the 1940s when Herbert Morrison, one of the 'founding fathers' of state ownership in the Attlee government, said that '...a public corporation gives us the best of both worlds' because it can '...combine modern business management with a proper degree of public accountability' (House of Commons 1946).

After a honeymoon period in the 1950s and early 1960s, public disillusionment with nationalisation grew. There were complaints of inefficiency, technological backwardness, lack of concern for consumers and poor industrial relations. Tensions between the Boards of the corporations and governments increased.

By the 1970s there was serious concern about the poor performance of Britain's nationalised industries, including utilities such as water, electricity, gas and telecommunications.[5]

After three decades of this poor performance, there were few who still subscribed to the idealised Morrisonian view of state corporations. Opinions about remedies varied, but the deficiencies of nationalisation were all too obvious.

A series of White Papers issued in 1961, 1967 and 1978 failed to bring about any improvement.[6] There were various attempts to impose some of the concepts of welfare economics on the industries – such as long run marginal cost pricing and test discount rates comparable to those used for low-risk private sector projects. Yet even these attempts foundered due to the sheer practical difficulties of implementing such ideas, and because of resistance from both politicians and the industries.

Most people would agree on some functions that must be performed or at least administered by government – defence and law and order, for instance. But, in the case of the British nationalised industries, governments strayed far outside these so-called 'public goods' into activities that are plainly commercial – where a product or service can be sold for a price to a willing buyer – and thus, there is no reason why the public sector should have replaced private enterprise in those activities.[7]

Problems inherent in nationalisation

Some of the problems that arose in the British nationalised corporations and that seem inherent in such a form of organisation are discussed below.

1. Ownership by no one and its consequences

One of the most serious ingrained problems in markets where there are state corporations is that citizens have virtually no means of influencing what those corporations do. 'Public' ownership in this sense is valueless because the 'owners' have no defined, transferable property rights in the organisation: in the well-known phrase, what is owned by everyone is perceived to be owned by no-one.[8] The 'agency' problem always exists when there is a divorce between an organisation's ownership and its management: the problem is maximised in the case of nationalised corporations. They have no shareholders other than government; they are immune to the pressures usually exerted by shareholders on management; and they cannot be taken over.

The owners of any company need means of monitoring and controlling the actions of the managers who are their agents: they do not want those managers to pursue their own interests but those of the owners. It is notoriously difficult, even under private ownership, to devise incentive structures (such as performance-related rewards) that align the interests of both owners and managers.

Nevertheless, shareholders in companies have the power of 'exit' as well as 'voice'. Complaints to managers or protests at Annual General Meetings may not be very effective, but the prospect of a plunging share price – as disgruntled shareholders 'exit' by selling their holdings in protest against underperforming managers – is a remarkably effective way of concentrating managers' minds. The wealth of managers who are shareholders will be reduced and, more important, the decline in the company's stock market value may make it a target for a potential bidder.

'Owners' of nationalised corporations lack the power of exit that private shareholders enjoy: they have no property rights to sell if they are dissatisfied with the state corporation's performance. Instead, they must rely on making their voices heard, principally through the politicians and civil servants who are the immediate principals of the state corporation management. Such indirect influence is highly unsatisfactory – unless one assumes that politicians and civil servants are altruistic, wise, well-informed individuals devoted to the interests of the community as a whole, and then further assumes that those interests can be discovered and pursued in some way other than through market processes (Robinson 2003).

Governments and government-appointed regulatory bodies cannot reasonably be assumed to be purely 'public-spirited'. As public choice theorists have pointed out (Buchanan 1978), one of the inconsistencies in mainstream economic theory is that it assumes people in the private sector pursue their own interests whereas people in the government sector pursue the public interest.

There seems no justification for this 'bifurcated man' assumption. More reasonably, one might assume that people are much the same wherever they work and that government sector employees are as likely to pursue self-interest as those in the private sector (Tullock 2000). Obviously, that assumption leads to quite different predictions of how those in the government sector will behave. In practice, it means not only do politicians and civil servants lack relevant information, they may have all manner of questionable objectives in mind for the corporations other than their being efficient and responsive to the wishes of citizens/customers.

2. Weak efficiency pressures

Given these agency problems, efficiency pressures on state-owned corporations are extremely weak. They are subject to monitoring by government departments but – in the absence of capital market comparisons – these departments have no way to determine how efficient the corporations are.

The problem is compounded if, as is often the case, the state corporation has a monopoly of the national product market. Private monopoly can be a problem but private companies that exploit their market power usually find that, in the course of time, rivals enter their market and compete away their profits. State monopolies are, however, protected by statute from entry. Consumers are captives, unable to exit from their existing supplier. There are no rivals to drive innovation, cost reduction, higher standards and lower prices. A government or a government-appointed regulator is unable to foresee what the outcome of a competitive market would have been: it has no relevant standard of comparison against which to judge the corporation and is forced to rely on unsatisfactory efficiency comparisons and efficiency audits.

3. Politicisation

Another serious issue in markets where there are state corporations is politicisation, which has implications for efficiency. Because politicians are likely to be held responsible for major, and sometimes minor decisions by state corporations, they tend constantly to interfere with decision-making. A common complaint from senior managers of the British nationalised corporations was that governments would not allow them to manage.

In the days of nationalised industries, governments would lean heavily on the nationalised corporations to pursue changing political objectives. They were both controlled and owned by the state. Nationalisation is a form of regulation but one without clear rules and predictable outcomes: the regulated company is usually subjected to backdoor pressure from politicians and civil servants. Before privatisation, the British nationalised industries were, at times, induced to keep their prices artificially low in order to make the general rate of inflation appear lower. At other times, they had to increase or to decrease investment, not according to the prospective rate of return on capital, but depending on the financial position of the government and whether it was seeking to boost or restrict the rate of economic growth.

Morrison believed that politicians and nationalised industry management could maintain an 'arm's length' relationship but, given the ill-defined responsibility governments had for the industries, the political interference that was so resented by management came not just at the 'macro' level as explained above, but in 'micro' detail. Managerial objectives were confused by doubts whether the industries should follow 'commercial' or 'public service' aims or simply do the bidding of the government of the day. By their actions, governments created severe regulatory uncertainty in the industries they controlled.

One consequence of nationalised ownership is that lobbying is rife. Management realises that their activities are affected at least as much by the actions of politicians and civil servants as by their own efforts to innovate and cut costs. Lobbying appears to be a relatively high-return activity into which corporate resources therefore inevitably flow, diverting scarce management resources away from innovation and efficiency improvement. Lobbying is, of course, present also in markets where there are no state corporations – but the relatively high returns which accrue to state corporations through lobbying activities means that it is innate in such organisations.

To summarise, in the case of monopoly state corporations, the absence of competitive market forces means that pressures on the corporations to increase efficiency and to pass gains on to consumers are very weak. Politicisation is rife and resources flow into lobbying. Attempts by governments and regulators to simulate the results of competitive pressures are a pale shadow of the real thing: in the absence of information from either capital markets or product markets, virtually all the facts required for meaningful efficiency comparisons are absent.

English, Welsh and Scottish water: problems of nationalised monopoly

The English and Welsh water industry was privatised in 1989. Ten large vertically-integrated water and sewerage companies were created in England and Wales (hereafter 'water companies') out of the previous river basin authorities, in addition to the 29 already privately-owned water-only companies. As with Britain's other privatised utilities, a regulatory office – the Office of Water Services (Ofwat) – was established to apply price cap controls, and water became subject to an environmental regulatory regime (now by the Environment Agency).

Privatisation has had numerous benefits. It has allowed businesses to raise capital without going to the government. The industry is regulated independently by Ofwat,[9] which has, as in other utilities, reduced the politicisation of decision-making that was a feature of state ownership. Incentive regulation has improved the efficiency of the water companies and regulation is generally recognised to be open and transparent (Mayer 2005). It can therefore be argued that water privatisation and regulation have been successful (Mayer 2005).

However, there are still many flaws in this privatised structure, in which a number of regional monopolies are supervised by an 'economic' regulator, and environmental and quality regulators. Up to now, there has been very little competition (and that only for a few very large customers), and it is not clear that a new regime under the 2003 Water Act will be effective in liberalising the non-household market (Robinson 2004).

The regulator operates primarily by using 'yardstick' or 'comparative' competition to compare different companies: as explained below, that is not a very satisfactory basis. Other problems arise because existing companies have some protection from takeover (ibid.). To avoid reducing the number of 'comparators', a constraint has been placed on 'water-to-water' mergers. A water company in England and Wales that wants to take over another water company is automatically referred to the Competition Commission, and the Commission must consider the effect of the merger on Ofwat's ability to make comparisons. Such mergers have generally not been allowed and the market for corporate control has therefore been constrained.[10] The result is perverse since the reason for instituting comparative competition is the absence of real competition. But comparative competition then leads to the merger restriction which means that, not only is product market competition absent, but the capital market – normally a discipline on inefficient managers – does not work properly either.

Furthermore, a significant feature of the industry is that, as well as 'economic' regulation, water companies are subject to detailed environmental and quality regulation, both from the Environment Agency in Britain and from the European Union, which forms a very important part of the supervisory system as a whole. Indeed, EU directives on water quality have been responsible for most of the price increases since privatisation (Helm and Rajah 1994; *Financial Times* 2004). In general, the industry seems to suffer from too much regulation.

Despite some of the problems experienced in England and Wales, the industry has made some advances since privatisation (Mayer 2005) whereas Scottish Water still exists in a setting similar to that which formerly characterised the British nationalised sector as a whole. The deficiencies of that sector, described above, are all too easily recognisable in the water regime in Scotland as it has operated so far. There is 'public' ownership, so there are no shareholders with property rights and no share price. There is monopoly: Scottish Water has so far been the sole provider of water and waste water services in Scotland, so consumers have no choice. There is politicisation: the Executive Board of Scottish Water answers to the Scottish Parliament.

As well as this direct political control, Scottish Water is – like the privatised English and Welsh companies – subject to an 'economic' regulator (WICS, up to the 2005 Act, and now, the Water Industry Commission) which regulates charges and service standards. However, in a significant difference from England and Wales, before the 2005 Act the regulator worked as an adviser to Ministers. SW is also, like the industry in England and Wales, subject to numerous quality, environmental and health and safety regulators – in SW's case, these are the Drinking Water Quality Regulator, the Scottish Environmental Protection Agency and the Health and Safety Executive.

Politicians have not been shy of involving themselves in the company's business, as one would expect given the statutory position of Scottish Water and previous experience of nationalised corporations. The Water Industry (Scotland) Act 2002 permitted Scottish Ministers to give guidance to WICS about how he should perform his functions and Ministers set out guidelines for Scottish Water. In February 2005, for example, the Minister provided objectives for the water industry in Scotland from 2006 to 2014.[11]

The objectives told Scottish Water that it should not increase charges by more than the inflation rate between 2006 and 2010. In addition, various objectives were established relating *inter alia* to drinking water quality, environmental improvement, sewer flooding prevention, connection of new homes and the rebalancing of charges in favour of businesses. Scottish Water must draft a business plan to show how it will comply with the Ministers' demands. Plainly, once Ministers have stated their objectives, they are bound to monitor Scottish Water's progress in achieving them and to exercise control when the company appears to be falling short. Thus the intervention in company decision-making that was formerly a common feature in British nationalised industries seems inevitable in the Scottish water industry.

The 'economic' regulator before the 2005 Act, WICS, lacked some of powers of Ofwat in England and Wales. WICS did not itself set charges as does Ofwat.[12] Its role was indirect, with principal duties including:

- to advise Ministers on the revenue required by Scottish Water to provide customers with a 'sustainable service' and to fund its investment programme;
- to consider and approve Scottish Water's annual scheme of charges (though with any disputes being referred to Ministers);
- to advise Ministers on Scottish Water's service standards and customer relations; and
- to advise Ministers, when requested, on a range of matters relating to Scottish Water's impact on customers.

Thus, its role was as Ministerial adviser rather than independent regulator. Furthermore, though there has been a move in Britain to remove competition policy from political control, before the 2005 Act disputes between WICS and SW were resolved by Ministers rather than being referred to the Competition Commission (as they would be in similar cases in England and Wales).

Although WICS said that there were safeguards for his independence of view and that he was not controlled by Ministers,[13] his position as adviser to Ministers was well out of line with the rest of the utility regime in Britain. That regime has three 'pillars' – regulatory offices independent of political control, competition promotion duties for the regulators and incentive-based price control of monopolies (Robinson and Marshall 2006). Before the 2005 Act, WICS lacked the first two of these, which are arguably the most important, and did not itself control the price cap.

Contemplating the Scottish water regime as a whole before the 2005 Act, it embodied most of the worst features of an old-style nationalised system in which politicians and civil servants attempt to run a major industry. Government controlled not just the general direction of the industry but also its charges and standards of service. The authors of that regime might have learned from decades of practical experience in Britain, as well as theoretical considerations, that the performance of nationalised corporations is not just an aberration from some much superior norm that can be achieved in the Scottish water industry.

Far from following the public interest, nationalised corporations have *inherent* undesirable characteristics: inefficiency, politicisation and disregard for the interests of their (captive) customers. Indeed, it is significant that WICS used the performance of the privatised industry in England and Wales as his standard of efficiency. One of WICS' objectives – to '...ensure that the level of customer service is on a par with the service delivered in England and Wales' (WICS 2004a, 13) – was an implicit admission that, in England and Wales, ownership, structure and regulatory regime combine to produce a superior performance.

Starting a revival? The 2005 Water Services Act

Some of the problems of the Scottish water industry, including the tensions in the relationship between WICS and Ministers, may be eased by the passage in February 2005 of the Water Services etc (Scotland) Act which, among other things, establishes a new Water Industry Commission which bears a closer resemblance to Ofwat. The Commission has powers to set charges (within policy guidelines from Ministers) and Scottish Water has a right of appeal to the Competition Commission against price determinations, thus bringing the Scottish regulatory system for water closer to that of England and Wales (Scottish Executive 2004; Water Services etc (Scotland) Bill 2005).

The new Act is both a recognition that up to now, the 2002 Act has created major problems in Scotland's water supply regime, and a first step along the road to correcting some of the errors of the past. It gives greater independence to the regulator and provides for limited competition for non–household consumers. But it is no more than an initial step. Fundamental difficulties remain – above all, that Scottish Water is still a nationalised corporation, subject to political pressures, with substantial monopoly power and poor incentives for management. Some specific problems – both of commission and of omission under the new regime – are discussed below.

The regulator gains considerably in independence from political control and the ability of Scottish Water to appeal against regulatory decisions to the Competition Commission is a step forward in reducing political influence. But, since Scottish Water remains in 'public' ownership, it is not clear to what extent the Scottish water industry has been freed from politicisation.

Although the new Water Industry Commission has the power to determine Scottish Water's charges, its freedom is bounded by constraints of 'principles' set out by Ministers about 'charge limits for different consumer groups' (Scottish Executive 2004, paras 3.5 and 3.6). In February 2005, the deputy environment Minister announced that the poorest households would be given a 25 per cent discount on their water charges. This will be paid for by abolishing the 25 per cent discount now enjoyed by those individuals who own more than one home. Such actions suggest that Ministers are pursuing distributional objectives through water pricing (*Scotsman* 2005).

Another problem is that in Scotland, SW does not charge consumers directly, thus reducing contact between supplier and customer. Charges are collected by local authorities and may thus impinge on their ability to collect council tax, increasing the risk that there will be political manipulation of water charges (*Edinburgh Evening News* 2005). Ministers also have a duty to provide Scottish Water with 'standards and objectives…in the provision of core services' (Scottish Executive 2004, para 3.4). Past experience of relations between Ministers and nationalised industries suggests that such provisions will still permit interference by Ministers who wish to do so.

Continued government ownership also means that there will be no efficiency pressures stemming from shareholders and that Scottish Water will remain outside the market for corporate control. The lack of capital market efficiency pressures is a serious matter, given the apparent inefficiency revealed by the regulator's studies.

Of course, if product market competition were to emerge, that would itself increase efficiency pressures and provide better incentives for managers, since competing companies would have an incentive to reduce their costs and to innovate. But the new Act goes so far as to prohibit competition for domestic consumers. Almost always and almost everywhere, politicians' bans on competition are an extremely bad idea.

The Act also rules out a common carriage system that might be another way to promote competition. Moreover, it gives Ministers a role in licensing entrants to the industry, permitting them to specify 'other factors' (other than those a regulator would normally take into account) in deciding whether an applicant is suitable – a provision which clearly could be abused.

An innovation in the legislation is the possibility of competition for larger consumers between entrants and a new 'arms-length' subsidiary of Scottish Water. SW would not be allowed to discriminate in favour of this subsidiary and against entrants. Licensed entrants would seek water supplies from Scottish Water which, if agreement was reached, would supply water to the customer through the 'public' supply system.

Unfortunately, however, this 'competitive' regime – like its counterpart in England and Wales, under the 2003 Water Act – seems likely to experience the problems that usually plague regulated access systems (Robinson 2004). Experience with British Gas, for instance, suggests that a considerable advantage is enjoyed by an incumbent which controls the pipeline network (as will SW) and from whom entrants are required to request a water supply. It is difficult to avoid discrimination against potential entrants, particularly since SW can plead that entry might jeopardise the performance of its statutory functions.

Moreover, experience in England and Wales suggests that bureaucratic delays in regulated access arrangements and delays in the resolution of complex disputes by the regulator are likely to put off prospective entrants. As the Monopolies and Mergers Commission (now the Competition Commission) remarked of the old British Gas regime, it was incapable of providing the 'necessary conditions for self-sustaining competition' (Monopolies and Mergers Commission 1993). Unlike British Gas, SW would have a separate subsidiary – but even so, the prospect of entering a market dominated by a nationalised competitor is unlikely to seem attractive unless the regulator adopts an open and determined pro-competitive stance.

If it is true that competition will not flourish under the provisions of the new Act, except perhaps at the margins of the industry, efficiency pressures on SW from product markets will continue to be weak. Since, as explained above, SW faces no shareholder efficiency pressures either, the prospects for improved efficiency under the new regime do not look good.

If Scottish Ministers leave the regulator alone, the Commission will have a better chance than in the past of stimulating efficiency improvements in SW. But, in the absence of any significant information from markets, the regulator will have very little information on which to base comparative efficiency studies. Presumably he will fall back on comparisons with the English and Welsh water companies, as he does now: the amalgamations in Scotland have suppressed possible Scottish 'comparators'.

Yet these comparisons are unlikely to be fruitful. As explained above, there has been very little attempt to liberalise the industry in England and Wales which therefore suffers from an intrusive and tightening regulatory system (Robinson 2004). Because there is little real competition among the regional monopolies, regulators have made do with so-called 'competition by comparison' ('yardstick competition') which is extremely unsatisfactory. There are serious difficulties in making useful efficiency comparisons within England and Wales because of differences in the conditions in which the companies operate. The econometric models used seem far too weak to standardise for varying conditions and to form the basis for comparisons which are used for price-setting (Robinson 2002). Given this inherent weakness, it stretches the system even farther beyond its proper bounds to try to include in the English and Welsh regime the Scottish water industry, where conditions are different again.

Privatisation and liberalisation

It is difficult to see how the Scottish water industry can be revived unless it is privatised. But privatisation should only be regarded as a necessary first step, an enabling measure. In itself, it is not sufficient. Market liberalisation is particularly important and should accompany privatisation.

Privatisation subjects privatised companies to the discipline of the capital market. However, it may not itself result in product market liberalisation and so there is no guarantee that efficiency gains from privatisation will be passed on to consumers. A number of British privatisations have simply transformed state monopolies into private monopolies, at least in the short term. In the case of the railways, for example, there is very little competition except for franchises (and then winning companies receive state subsidies). As already explained, the water industry in England and Wales, though privatised, is divided into a number of regional monopolies and hardly any competition exists. British Gas was privatised whole in 1986 and, for a number of years after privatisation, had a virtual monopoly because it owned the pipeline system which others had to use, and it had most of available gas from the North Sea tied up in long term contracts.

The advantage of product market liberalisation is that it sets in motion competitive processes which add to the efficiency pressures stemming from the capital market and, crucially, it passes on the benefits to consumers. Competitive markets give the power of exit to consumers, as well as to shareholders. Thus there is constant pressure on producers to provide combinations of lower prices and higher standards which appeal to consumers, and which their competitors then try to emulate. In other words, there is a race to the top.

This kind of market process is what Adam Smith and later classical economists meant by competition – a process of dynamic change in which the *status quo* is constantly disturbed by entrepreneurs who are looking for better ways to produce goods and services.[14] Free entry is the key. That means not just removing statutory monopolies but establishing economic conditions in which competition can flourish. If competitors can enter the market, incumbents (unlike monopoly nationalised corporations) cannot ignore them but must respond to the prices and service standards they offer.

Twenty-five years ago, a question that might have been asked about liberalisation of the markets of nationalised corporations was – is competition feasible and, if feasible, is it desirable? Many people regarded these industries as 'natural monopolies' where efficiency dictated that there should be only one supplier. However, theoretical advances, now backed by substantial practical experience, show the natural monopoly argument to have been largely false. Some of the nationalised industries, such as coal-mining and the airlines, were *unnatural* monopolies: that is, they were state artefacts where there were no efficiency advantages from sole ownership (Robinson and Marshall 1985). In other cases, in the utilities, there are natural monopoly elements, but large parts of the industries are potentially competitive and there is no economic reason why they should not be liberalised.

A recent insight about the nature of 'network' utilities is that the traditional gas, water, electric or telephone utility consists of a network of pipes or wires which is (given existing technology) a natural monopoly, yet other activities such as production, storage and supply to consumers are potentially competitive.[15] There are considerable advantages to consumers in having actual competition introduced into these latter areas. For example, the good can be produced at the wholesale level by a number of rival companies, thus keeping down production costs and promoting innovation;

there can be competition in storage, in meter provision and in meter reading; and, at the final stage, of supplying customers, rival companies can compete in terms of price and service. Consumers have a choice of suppliers, now that they are no longer the captives of a nationalised monopoly, and so they gain the ability to switch to a better offer, if one is available.

An essential part of this scheme of liberalisation is that the rest of the industry, the natural monopoly network, should be separated (preferably in a separate company). Separating the network and regulating it as a natural monopoly allows competition to reign in the rest of the industry, minimising the unsatisfactory business of regulation and concentrating it on the sector where, at present, it appears unavoidable. This process has been carried to its logical end in gas and electricity where there is competition to generate electricity and to produce gas and, after the product has passed through the network, there is competition to supply gas and electricity at retail and wholesale levels.

Regulation of the network is necessary because, unlike the competitive areas of the utilities where supplier rivalry protects consumers, some specific consumer protection against exploitation by the monopoly, including discrimination against potential entrants, must be established. In the British utilities, the usual form of protection has been by an independent regulatory office (Ofgem, Ofcom, Ofwat, etc.) which uses an RPI-x price control, among other devices, to keep network charges within bounds.[16]

Competitive parts of the industry are not subject to price controls (except for an interim period in which competition is being established), though the regulator uses his or her competition-promotion duty and duties under the Competition Act 1998 to keep a general watch on them. The Competition Commission stands ready in the background to take action if necessary. In the sectors where this scheme has been applied – in which competition is introduced where possible and regulation is used where there is no immediate prospect of competition – it appears to have worked well. In the energy utilities, in particular, where the regulators have been most assiduous in promoting competition, competition flourishes in both wholesale and retail markets.

What to do

The present scheme of ownership, organisation and regulation of the water industry in Scotland shows virtually no sign of any application of economic principles nor does it recognise the insights that have already been gained from utility privatisation and regulation schemes. Only political expediency and bureaucratic convenience (and perhaps an emotional attachment to 'public' ownership, despite all the failings described above) can explain the original scheme and its modifications so far.

Under the new arrangements the industry remains essentially a nationalised monopoly (though there may be some competition at the margin for some large consumers). Thus, despite the welcome granting of greater independence to the regulator, it is not clear that the regulator and Scottish Water have escaped political control. If politicians did not wish to interfere, there would have been no reason to keep the industry nationalised.

Efficiency pressures are muted. The regulator may well continue to 'shadow' the England and Wales industry in an attempt to bring the Scottish industry up to the apparently much higher England and Wales standards. But, given the existing deficiencies of comparative competition in England and Wales, shadowing that system is most unsatisfactory. Indeed, it is paradoxical that an industry which has deliberately been kept nationalised should implicitly accept that it is inherently inferior to the privatised system in England and Wales by trying to model itself on privatised companies to obtain the benefits of their better standards of service.

The key objectives of reform should be to find ways to promote efficiency in the Scottish water industry and to pass those gains on to consumers. Experience elsewhere indicates that under nationalised monopoly these objectives are most unlikely to be achieved. Privatisation (to bring capital market disciplines) and liberalisation (to bring product market disciplines) are required. Reformers should bear in mind that water supply is, in principle, similar to the supply of gas and electricity, where competition now reigns over large parts of the industries. Production (extraction) and storage are dispersed and are potentially competitive; meter provision and meter reading are potentially competitive; the supply of water to consumers is also potentially competitive. Indeed, as in gas and electricity, the only 'natural monopoly' activity that requires regulation is the transport of water by pipeline.

More specifically, the following steps need to be taken.

First, though the new regulatory regime under the 2005 legislation is an improvement on its predecessor, there are lingering doubts about possible political interference with the regulator and with Scottish Water. To reduce regulatory uncertainty, these doubts should be dispelled by privatisation by public flotation. Even though such interference will not necessarily cease after privatisation, it will become significantly more difficult. Privatisation would also bring efficiency pressures from shareholders.

Second, entry to the industry should be made as easy as possible in the interests of stimulating competition. Licences to supply water should be freely available, subject only to the regulator being satisfied with the applicant's ability to perform the necessary functions. In contrast, a provision in the legislation that allows a role for Ministers to specify 'other factors' that would determine whether or not applicants should be granted licences should be eliminated.

Third, if the proposals to introduce competition for non-household customers are to have any chance of success, the regulator must apply a vigorous pro-competition policy to ensure that entry to the market occurs. Liberalisation of the market is so important that the regulator should be given a specific duty to promote competition.

Fourth, the prohibition on competition in the household market, which provides Scottish Water with millions of captive customers, should be removed. Household competition is not an imminent prospect – competitive supply to larger customers will be easier to introduce and is likely to come first. However, it would be most unwise to rule households out, given that in some British utilities (notably gas and electricity) households have been major beneficiaries of competition.

Fifth, the disconnection between the industry and its customers that occurs because charges are collected by local authorities should be ended. Customers should be billed by their supplier so it is clear who is responsible for supply and associated services.

Sixth, when Scottish Water is privatised, water pipelines should be separated from the rest of the company, not in a subsidiary but in a separate private, regulated company. A separate pipeline company is an important competition-promoting device. So long as Scottish Water controls pipeline access, entry is likely to be limited. A separate company, however, would have a powerful business interest in transporting water for all customers and would ensure there would be no discrimination against entrants.

Notes

1. I have received many helpful comments on a draft of this paper from Dr Eileen Marshall and a number of anonymous referees, none of whom has any responsibility for its conclusions. This paper is an edited version of 'Reviving the Scottish Water Industry', published by the Policy Institute (Scotland), Series: Economy No.9, March 2005 and also draws on observations made in Robinson (2004).
2. To keep the chapter reasonably brief it omits two issues that are closely related to its subject, but which merit separate examination. It discusses the supply of water only and does not deal with sewerage. Second, it does not consider, in any detail, environmental and quality regulation of the water industry. This form of regulation, driven primarily by directives from Brussels that aim to 'improve' water quality even though its authors have no idea how much consumers would be willing to pay for water of different qualities, is increasingly intrusive and needs reconsideration (see Robinson 2004).

3. Most regulatory bodies for British utilities were established after privatization. Another corporation which is still nationalized but has a regulator is the Post Office.
4. Explanatory Notes on the legislation can be found at www.scottish.parliament.uk/business/bills
5. See, for example, Heald (1980).
6. "The Financial and Economic Obligations of Nationalised Industries" (Cmnd.1337), "Nationalised industries. A review of economic and financial objectives" (Cmnd.3437) and "The Nationalised Industries" (Cmnd.7131), respectively.
7. Pure public goods are those where it is not possible to exclude people from their supply and where consumption is non-rivalrous (supply to one does not reduce supply to another, so the marginal cost is zero). Because all the benefits of these goods are 'externalities' private suppliers cannot appropriate any benefits and so will not be willing to supply. In practice, there are very few pure public goods though a number of goods and services have some public good characteristics. Even classic public goods such as law and order and defence do not have to be financed and supplied entirely by the state: sometimes voluntary collective action is possible and so is contracting out of service provision.
8. The problems of 'public' ownership are explained in Robinson (2003).
9. The 2003 Water Act establishes a regulatory authority, the Water Services Regulation Authority, as has been done in other utilities, where a board rather than a single regulator is now the norm.
10. There have been a number of takeovers of water companies by companies not already operating water companies in England and Wales.
11. The objectives are listed on Scottish Water's website www.scottishwater.co.uk
12. Both Ofwat and WICS are, of course, constrained in their actions by the views of the environmental and quality regulators
13. *Role of the Water Industry Commissioner for Scotland*, www.watercommissioner.co.uk
14. For an explanation see Blaug (1987).
15. Due primarily to the late Professor Michael Beesley. See Beesley (1997).

16. RPI, the retail price index (a measure of inflation), minus x, which is a number devised by the regulator, intended to give an incentive to improve efficiency.

References

Beesley, M.E. (1997). *Privatisation, Regulation and Deregulation.* London: Routledge, second edition.

Blaug, M. (1987). "Classical Economics", in J.Eatwell, M.Milgate and P.Newman (eds), *The New Palgrave – A Dictionary of Economics,* vol.1. Hampshire, UK: Palgrave Macmillan.

Buchanan, J. (1978). *The Economics of Politics.* IEA Readings 18, London: Institute of Economic Affairs.

Edinburgh Evening News (2005)."City leaders want to pull plug on collection of water rates". 19 January.

Financial Times (2004). "Limits on Water Spending Urged". 22 January.

Heald, D. (1980). "The Economic and Financial Control of UK Nationalised Industries". *Economic Journal,* June.

Helm, Dieter and Najma Rajah (1994). "Water Regulation: The Periodic Review". *Fiscal Studies,* Vol.15 No.2, May.

House of Commons (1946). Hansard 6 May, Cols 604–5.

Mayer, C. (2005). "Commitment and Control in Regulation: The Future of Regulation in Water", in Colin Robinson (ed.), *Governments, Competition and Utility Regulation,* Cheltenham, UK: Edward Elgar.

Monopolies and Mergers Commission (1993). *Gas and British Gas plc,* Cmnd.2314–2317, Vol.1, para 1.6. London: HMSO.

Robinson, C. (2002). "Moving to a Competitive Market in Water", in Robinson (ed.) *Utility Regulation and Competition Policy.* Cheltenham, UK: Edward Elgar.

Robinson, C. (2003). "Privatisation: analysing the benefits", in David Parker and David Saal (eds.), *International Handbook on Privatization.* Cheltenham, UK: Edward Elgar.

Robinson, C. (2004). ''Water privatization: too much regulation?'. *Economic Affairs,* Vol.24, no.3, September.

Robinson, C. and Eileen Marshall (1985). *Can Coal be Saved?,* Institute of Economic Affairs, Hobart Paper 105.

Robinson, C. and Eileen Marshall (2006). "Regulation of Energy: Issues and Pitfalls", in David Parker and Michael Crew (eds.), *International Handbook of Regulation*. London: Edward Elgar, forthcoming.

Scottish Executive (2004). Letter from Ross Finnie to WICS. 26 May. Online: http://www.watercommissioner.co.uk/Documents/src2006/FD/appendices/Appendix2.pdf

Scotsman (2004). "Scots water service 'is the worst in the UK'". 19 November.

Scotsman (2005). "Ministers limit water charge rises to rate of inflation". 10 February.

Tullock, G. (2000). "The Theory of Public Choice", in G. Tullock, A. Seldon and G. Brady, *Government: Whose Obedient Servant?*, IEA Readings 51, London: Institute of Economic Affairs.

Water Industry Commissioner for Scotland (WICS)(2004a). *Our work in regulating the Scottish water industry: Setting out a clear framework for the Strategic Review of Charges 2006–10*. Volume 1, July. Online: http://www.watercommissioner.co.uk/Documents/src2006/Our%20work%20in%20regulating%20the%20Scottish%20water%20industry.pdf

—— (2004b). *Customer Service Report, 2002–03: Scottish Water*. November, chapter 6. Online: http://www.watercommissioner.co.uk/documents/final.pdf

—— (2004c). *Our work in regulating the Scottish water industry: the scope for capital investment efficiency*. Volume 5, December. Online: http://www.watercommissioner.co.uk/Documents/src2006/volume5/VOL5_COMPLETE.PDF

Water Services etc (Scotland) Bill (2005). Explanatory Memorandum. Online: http://www.opsi.gov.uk/si/si2005/draft/em/uksidem_0110697529_en.pdf

The Sustainable Development Network

www.sdnetwork.net

The Sustainable Development Network is a coalition of individuals and non-governmental organizations who believe that sustainable development is about empowering people, promoting progress, eliminating poverty and achieving environmental protection through the institutions of the free society.

The SDN promotes the view that sustainable development can only be achieved with evolutionary institutions that harness human initiative, including property rights, contracts, the rule of law, open markets and open trade, and accountable, transparent government.

The SDN develops and promotes policy solutions to achieve real sustainable development, by enabling people to improve their own wellbeing (and that of others), pursue their own goals, and protect the environment, without bureaucratic intervention (whether local, national and international). Members of the SDN generate policy materials, participate in global fora and engage with the news media to encourage public discussion of this institutional approach to sustainable development.

SDN Members

Ag Bio Foundation, USA
www.agbioworld.org

Africa Fighting Malaria, South Africa
www.fightingmalaria.org

Alternate Solutions Institute, Pakistan
www.asinstitute.org

ARCH-Vahini, Gujarat, India

Asociación de Consumidores Libres, Costa Rica
www.consumidoreslibres.org

Association for Liberal Thinking, Turkey
www.liberal-dt.org.tr

CEDICE, Venezuela
www.cedice.org

Centre for Civil Society, India
www.ccsindia.org

Centre for Environmental Studies
Liberalni Institute
Prague, Czech Republic
www.libinst.cz

Centre for New Europe, Belgium
www.cne-network.org

CEPPRO, Paraguay

China Sustainable Development Research Center, Capital University of Business & Economics

Circulo Liberal, Uruguay
www.circuloliberal.org

Community Resources Institute, Kenya

Fundacion Atlas 1853, Argentina
www.atlas.org.ar

Fundacion Libertad, Panama
www.fundacionlibertad.org.pa

Free Market Foundation, South Africa
www.freemarketfoundation.com

Instituto Ecuatoriano de Economía Política, Ecuador
www.ieep.org.ec

International Policy Network, UK
www.policynetwork.net

Imani – the Centre for Humane Education, Ghana
www.imanighana.org

INLAP, Costa Rica
www.inlap.org

Instituto de Libre Empresa, Peru
www.ileperu.org

Instituto Liberdade, Brazil
www.il-rs.com.br

Instituto Libertad y Progreso, Colombia
www.ilyp.net

Institute of Public Affairs, Australia
www.ipa.org.au

Inter-Region Economic Network, Kenya
www.irenkenya.org

Institute for Public Policy Analysis, Nigeria
www.ippanigeria.org

Libertad y Desarrollo, Chile
www.lyd.cl

Liberty Institute, India
www.libertyindia.org

Lion Rock Institute, Hong Kong
www.lionrockinstitute.org

Manushi, India
indiatogether.org/manushi

Research Center for Entrepreneurship Development, Vietnam
www.rced.com.vn

Zambia Institute for Public Policy Analysis
www.zippazambia.org